Student Solutions Manual

for

Bettelheim, Brown, Campbell, and Farrell's

Introduction to General, Organic, and Biochemistry

Eighth Edition

Mark S. Erickson
Hartwick College

Shawn O. Farrell
Olympic Training Center

Courtney A. Farrell

Australia • Brazil • Canada • Mexico • Singapore • Spain • United Kingdom • United States

Printed in the United States of America
1 2 3 4 5 6 7 10 09 08 07 06

Printer: Thomson/West

0-495-01421-4

Cover Image: Thomas G. Barnes @ USDA-NRCS PLANTS Database / Barnes, T.G. & S.W. Francis. 2004. *Wildflowers and Ferns of Kentucky.* University Press of Kentucky.

For more information about our products, contact us at:
Thomson Learning Academic Resource Center
1-800-423-0563

For permission to use material from this text or product, submit a request online at
http://www.thomsonrights.com.
Any additional questions about permissions can be submitted by email to **thomsonrights@thomson.com.**

Thomson Higher Education
10 Davis Drive
Belmont, CA 94002-3098
USA

Student Solutions Manual
for
Introduction to General, Organic, and Biochemistry,
Eighth Edition

Table of Contents

Introduction to Organic and Biochemistry, Fifth Edition

Table of Contents for Organic and Biochemistry Users

Chapter 1 Matter, Energy, and Measurement

1.1 Multiplication: (a) 4.68×10^{5} (b) 2.8×10^{-15}
Division: (a) 1.94×10^{18} (b) 1.36×10^{5}

1.3 $$241\ \cancel{lb}\left(\frac{453.6\ \cancel{g}}{1\ \cancel{lb}}\right)\left(\frac{1\ kg}{1000\ \cancel{g}}\right) = 109\ kg$$

1.5 $$\frac{332\ \cancel{m}}{\cancel{s}}\left(\frac{1\ \cancel{km}}{1000\ \cancel{m}}\right)\left(\frac{1\ mi}{1.609\ \cancel{km}}\right)\left(\frac{60\ \cancel{s}}{1\ \cancel{min}}\right)\left(\frac{60\ \cancel{min}}{1\ hr}\right) = 743\ mi/hr$$

1.7 $$d = m/V = \frac{56.8\ g}{23.4\ mL} = 2.43\ g/mL$$

1.9 $$\text{Amount of heat} = SH \times m \times (T_2\text{-}T_1) = \frac{1.0\ cal}{\cancel{g}\cdot\cancel{^{\circ}C}}(731\ \cancel{g})(74-8)\cancel{^{\circ}C} = 4.8 \times 10^{4}\ cal$$

1.11 $$SH = \frac{\text{Amount of heat}}{m \times (T_2\text{-}T_1)} = \frac{88.2\ cal}{(13.4\ g)(176\text{-}23)^{\circ}C} = 0.0430\ cal / g\cdot{}^{\circ}C$$

1.13 (a) Matter is anything that has mass and takes up space.
(b) Chemistry is the science that studies matter.

1.15 Dr. X's claim that the extract cured diabetes would be classified as a (c) hypothesis. No evidence had been provided to prove or disprove the claim.

1.17 (a) 3.51×10^{-1} (b) 6.021×10^{2} (c) 1.28×10^{-4} (d) 6.28122×10^{5}

1.19 (a) 6.48×10^{7} (b) 1.6×10^{5} (c) 4.69×10^{5} (d) 2.8×10^{-15}

1.21 (a) 1.3×10^{5} (b) 9.40×10^{4} (c) 5.137×10^{-3}

1.23 4.45×10^{6}

1.25 (a) 2 (b) 5 (c) 5 (d) 5
(e) 3 (f) 3 (g) 2

1.27 (a) 92 (b) 7.3 (c) 0.68 (d) 0.0032 (e) 5.9

1.29 (a) 1.53 (b) 2.2 (c) 0.00048

1.31 Answers rounded to two significant figures:

$$20\ \cancel{ks}\left(\frac{1000\ \cancel{s}}{1\ \cancel{ks}}\right)\left(\frac{1\ min}{60\ s}\right) = 330\ min$$

$$20\ \cancel{ks}\left(\frac{1000\ \cancel{s}}{1\ \cancel{ks}}\right)\left(\frac{1\ \cancel{min}}{60\ \cancel{s}}\right)\left(\frac{1\ hr}{60\ \cancel{min}}\right) = 5.6\ hr$$

1.33 (a) 20 mm (b) 1 inch (c) 1 mile

1.35 (b) Weight would change slightly. Mass is independent of location, but weight is a force exerted by a body influenced by gravity. The influence of the Earth's gravity decreases with increasing distance from sea level.

1.37 Temperature conversions: $°F = \frac{9}{5}\ °C + 32$ and $K = 273 + °C$

(a) $°F = \frac{9}{5}\ 25°C + 32 = \underline{77°F}$ and $K = 273 + 25°C = \underline{298\ K}$

(b) $°F = \frac{9}{5}\ 40°C + 32 = \underline{104°F}$ and $K = 273 + 40°C = \underline{313\ K}$

(c) $°F = \frac{9}{5}\ 250°C + 32 = \underline{482°F}$ and $K = 273 + 250°C = \underline{523\ K}$

(d) $°F = \frac{9}{5}\ (-273)°C + 32 = \underline{-459°F}$ and $K = 273 + (-273)°C = \underline{0\ K}$

1.39 Metric unit conversions:

(a) $96.4\ \cancel{mL}\left(\frac{1\ L}{1000\ \cancel{mL}}\right) = 0.0964\ L$

(b) $275\ \cancel{mm}\left(\frac{1\ cm}{10\ \cancel{mm}}\right) = 27.5\ cm$

(c) $45.7\ \cancel{kg}\left(\frac{1000\ g}{1\ \cancel{kg}}\right) = 4.57 \times 10^4\ g$

(d) $475\ \cancel{cm}\left(\frac{1\ m}{100\ \cancel{cm}}\right) = 4.75\ m$

(e) $21.64\ \cancel{cc}\left(\frac{1\ mL}{1\ \cancel{cc}}\right) = 21.64\ mL$

(f) $3.29\ \cancel{L}\left(\frac{1000\ cc}{1\ \cancel{L}}\right) = 3.29 \times 10^3\ cc$

(g) $0.044\ \cancel{L}\left(\frac{1000\ mL}{1\ \cancel{L}}\right) = 44\ mL$

(h) $711\ \cancel{g}\left(\frac{1\ kg}{1000\ \cancel{g}}\right) = 0.711\ kg$

(i) $63.7\ \cancel{mL}\left(\frac{1\ cc}{1\ \cancel{mL}}\right) = 63.7\ cc$

(j) $0.073\ \cancel{kg}\left(\frac{1000\ g}{1\ \cancel{kg}}\right)\left(\frac{1000\ mg}{1\ g}\right) = 7.3 \times 10^4\ mg$

(k) $83.4\ \cancel{m}\left(\frac{1000\ mm}{1\ \cancel{m}}\right) = 8.34 \times 10^4\ mm$

(l) $361\ \cancel{mg}\left(\frac{1\ g}{1000\ \cancel{mg}}\right) = 0.361\ g$

1.41 Speed limit $= \left(\frac{80\ \cancel{km}}{hr}\right)\left(\frac{1\ mi}{1.609\ \cancel{km}}\right) = 50\ mph$

1.43 Liquids and solids have definite volumes.

1.45 Melting is a physical change, not a chemical change; therefore, when a substance melts from a solid to a liquid, its chemical nature does not change.

1.47 Manganese (d = 7.21 g/mL) is more dense than the liquid (d = 2.15 g/mL) therefore it will sink. Sodium acetate (d = 1.528 g/mL) is less dense than the liquid; therefore it will float on the liquid. Calcium chloride (d = 2.15 g/mL) has a density equal to that of the liquid; therefore it will stay in the middle of the liquid.

1.49 $d_{urine\ sample} = m/V = \left(\frac{342.6\ g}{335.0\ \cancel{cc}}\right)\left(\frac{1\ \cancel{cc}}{1\ mL}\right) = 1.023\ g/mL$

1.51 Water (d = 1.0 g/cc) will be the top layer in the mixture because its density is lower than dichloromethane (d = 1.33 g/cc).

1.53 Water reaches its maximum density at 4°C; therefore, by warming the water from 2°C to 4°C, the lighter crystals will float on the water with an increased density at 4°C.

1.55 While driving your car, the car's kinetic energy (energy of motion) is converted by the alternator to electrical energy, which charges the battery, storing potential energy.

1.57 $SH_{unknown} = \frac{Heat}{m \times (T_2 - T_1)}$

$SH_{unknown} = \frac{2750\ cal}{168\ g\ (74^\circ - 26^\circ)^\circ C} = 0.34\ cal / g \cdot {}^\circ C$

1.59 Drug dose$_{135lb}$ $= 135 \text{ lb man} \left(\dfrac{445 \text{ mg drug}}{180 \text{ lb man}} \right) = 334 \text{ mg drug}$

1.61 The body first reacts to hypothermia by shivering. Further temperature lowering results in unconsciousness and later, followed by death.

1.63 Methanol would make a more effective cold compress because its higher specific heat allows it to retain the heat longer per mass unit.

1.65 $d_{brain} = \dfrac{1 \text{ lb}}{620 \text{ mL}} \left(\dfrac{453.6 \text{ g}}{1 \text{ lb}} \right) = 0.732 \text{ g/mL}$

$$\text{Specific gravity}_{brain} = \frac{d_{brain}}{d_{H_2O}} = \frac{0.732 \text{ g/mL}}{1 \text{ g/mL}} = 0.732$$

1.67 (a) Potential energy (b) Kinetic energy (c) Potential energy
(d) Kinetic energy (e) Kinetic energy

1.69 Convert European car's fuel efficiency of 22 km/L into mi/gal, then compare:

$$\text{Fuel efficiency}_{European} = \frac{22 \text{ km}}{\text{L}} \left(\frac{1 \text{ mi}}{1.609 \text{ km}} \right) \left(\frac{3.785 \text{ L}}{1 \text{ gal}} \right) = 52 \text{ mi/gal}$$

The European car is more fuel efficient by 12 miles per gallon

1.71 Shivering generates kinetic energy.

1.73 Convert each quantity into a common unit (grams): (a) is the largest and (d) is the smallest.

(a) 41 g

(b) $3 \times 10^3 \text{ mg} \left(\dfrac{1 \text{ g}}{1000 \text{ mg}} \right) = 3 \text{ g}$

(c) $8.2 \times 10^6 \text{ µg} \left(\dfrac{1 \text{ g}}{10^6 \text{ µg}} \right) = 8.2 \text{ g}$

(d) $4.1310 \times 10^{-8} \text{ kg} \left(\dfrac{1000 \text{ g}}{1 \text{ kg}} \right) = 4.1310 \times 10^{-5} \text{ g}$

1.75 Travel time = $1490\ \text{mi}\left(\frac{1.609\ \text{km}}{1\ \text{mi}}\right)\left(\frac{1\ \text{hr}}{220\ \text{km}}\right) = 10.9\ \text{hr}$

1.77 SH (water) = 1.000 cal/g · °C = 4.184 J/g · °C
SH (heavy water) = 4.217 J/g · °C
Heat = SH x m x ΔT: According to the equation, the heat required to raise the temperature of a substance is directly proportional to the specific heat of that substance. Heavy water, having the higher specific heat, will require more energy to heat 10 g by 10°C.

1.79 1.00 mL of butter = 0.860 g and 1.00 mL of sand = 2.28 g

(a) $d_{\text{mixture}} = \frac{3.14\ \text{g mixture}}{2.00\ \text{mL}} = 1.57\ \text{g/mL}$

(b) First calculate the volumes of sand and butter, which totals 1.60 mL

$$V_{\text{sand}} = 1.00\ \text{g sand}\left(\frac{1.00\ \text{mL sand}}{2.28\ \text{g sand}}\right) = 0.439\ \text{mL}$$

$$V_{\text{butter}} = 1.00\ \text{g butter}\left(\frac{1.00\ \text{mL butter}}{0.860\ \text{g butter}}\right) = 1.16\ \text{mL}$$

$$d_{\text{mixture}} = \frac{2.00\ \text{g}}{1.60\ \text{mL}} = 1.25\ \text{g/mL}$$

1.81 The final answer will be reported to two significant digits because the temperature is reported in two significant figures (the least accurate of the quantities used in the calculation)

$$SH_{\text{unk}} = \frac{\text{Heat}}{m \times (T_2 - T_1)} = \frac{3.200\ \text{kcal}}{(92.15\ \text{g})(45°\text{C})} = 7.7 \times 10^{-4}\ \text{kcal/g}\cdot°\text{C} = 0.77\ \text{cal/g}\cdot°\text{C}$$

1.83 $T_2 = \frac{\text{Heat}}{SH \times m} + T_1$

$$T_2 = \frac{60.0\ \text{J}}{(10.0\ \text{g})(1.339\ \text{J/g}\cdot°\text{C})} + 20.0°\text{C} = 24.5°\text{C}$$

1.85 Quantities of solids are most easily measured by their masses; therefore, the mass of urea is measured using a balance. Quantities of liquids are easily measured by their volumes and masses. The volume of pure ethanol can be measured using a volumetric pipette or graduated cylinder. The advantage of measuring the volume of ethanol is that the mass of ethanol can be calculated using its volume and density.

1.87 New medications are going to mimic natural molecules involved in biochemical processes, therefore, the new medicines would be expected to have similar characteristics to the natural biomolecules.

1.89 The contaminate can be removed from water using a technique called a liquid-liquid extraction. By adding diethyl ether to the aqueous sample in a separatory funnel, two layers are formed, with the less dense water-immiscible diethyl ether floating on top of the water layer. After mixing the two layers, the contaminant will dissolve in the upper diethyl ether layer. The lower aqueous layer is removed and the diethyl ether evaporated to isolate the contaminant.

2.1 (a) $NaClO_3$ (b) AlF_3

2.3 (a) The element has 15 protons, making it phosphorus (P); its symbol is $^{31}_{15}P$.

(b) The element has 86 protons, making it radon (Rn); its symbol is $^{222}_{86}Rn$.

2.5 The atomic number of iodine (I) is 53. The number of neutrons in each isotope is 125 - 53 = 72 for iodine-125 and 131 - 53 = 78 for iodine-131. The symbols for these two isotopes are $^{125}_{53}I$ and $^{131}_{53}I$.

2.7 This element has 13 electrons and, therefore, 13 protons. The element with atomic number 13 is Aluminum (Al).

$\dot{Al}:$

2.9 (a) Oxygen - an element
(b) Table salt - a compound
(c) Sea water - a mixture
(d) Wine - a mixture
(e) Air - a mixture
(f) Silver - an element
(g) Diamond - an element
(h) A pebble - a mixture
(i) Gasoline - a mixture
(j) Milk - a mixture
(k) Carbon dioxide - a compound
(l) Bronze - a mixture

2.11 Given here is the element, its symbol, and its atomic number:
(a) Bohrium (Bh, 107)
(b) Curium (Cm, 96)
(c) Einsteinium (Es, 99)
(d) Fermium (Fm, 100)
(e) Lawrencium (Lr, 103)
(f) Meitnerium (Mt, 109)
(g) Mendelevium (Md, 101)
(h) Nobelium (No, 102)
(i) Rutherfordium (Rf, 104)
(j) Seaborgium (Sg, 106)

2.13 The four elements named for planets are mercury (Hg, 80), uranium (U, 92) neptunium (Np, 93), and plutonium (Pu, 94).

2.15 (a) $NaHCO_3$ (b) C_2H_6O (c) $KMnO_4$

2.17 The law of conservation of mass

2.19 Mass percent of H and O in:
H_2O: 18.0 g/mol H: 11.2% O: 88.8%
H_2O_2: 34.0 g/mol H: 5.93% O: 94.1%

2.21 The statement is true in the sense that the number of protons (the atomic number) determines the identity of the element.

2.23 (a) The element with 22 protons is titanium (Ti).
(b) The element with 76 protons is osmium (Os).
(c) The element with 34 protons is selenium (Se).
(d) The element with 94 protons is plutonium (Pu).

2.25 Each would still be the same element because the number of protons has not changed.

2.27 Radon (Rn) has an atomic number of 86, so each isotope has 86 protons. The number of neutrons is mass number - atomic number.
(a) Radon-210 has 210 - 86 = 124 neutrons
(b) Radon-218 has 218 - 86 = 132 neutrons
(c) Radon-222 has 222 - 86 = 136 neutrons

2.29 Two more neutrons: tin-120
Three more neutrons: tin-121
Four more neutrons: tin 124

2.31 (a) An ion is an atom with an unequal number of protons and electrons.
(b) Isotopes are atoms with the same number of protons in their nuclei but a different number of neutrons.

2.33 Rounded to three significant figures, the calculated value is 12.0 amu. The value given in the Periodic Table is 12.011 amu.

$$\left(\frac{98.90}{100} \times 12.000 \text{ amu}\right) + \left(\frac{1.10}{100} \times 13.000 \text{ amu}\right) = 12.011 \text{ amu}$$

2.35 Carbon-11 has 6 protons, 6 electrons, and 5 neutrons.

2.37 Americium-241 (Am) has atomic number 95. This isotope has 91 protons, 91 electrons and 241 - 95 = 146 neutrons.

2.39 In period 3, there are three metals (Na, Mg, and Al), one metalloid (Si) and four nonmetals (P, S, Cl, and Ar).

2.41 Periods 1 - 3 contain more nonmetals than metals. Periods 4 -7 contain more metals than nonmetals.

2.43 Palladium (Pd), cobalt (Co), and chromium (Cr) are transition elements. Cerium (Ce) is an inner transition element and K and Br are main group elements.

2.45 (a) Argon is a nonmetal
(b) Boron is a metalloid
(c) Lead is a metal
(d) Arsenic is a metalloid
(e) Potassium is a metal
(f) Silicon is a metalloid
(g) Iodine is a nonmetal
(h) Antimony is a metalloid
(i) Vanadium is a metal
(j) Sulfur is a nonmetal
(k) Nitrogen is a nonmetal

2.47 The group number tells the number of electrons in the valence shell of the element.

2.49 (a) Li(3): $1s^22s^1$
(b) Ne(10): $1s^22s^22p^6$
(c) Be(4): $1s^22s^2$
(d) C(6): $1s^22s^22p^2$
(e) Mg(12): $1s^22s^22p^63s^2$

2.51 (a) He(2): $1s^2$
(b) Na(11): $1s^22s^22p^63s^1$
(c) Cl(17): $1s^22s^22p^63s^23p^5$
(d) P(15): $1s^22s^22p^63s^23p^3$
(e) H(1): $1s^1$

2.53 In (a), (b), and (c); the outer-shell electron configurations are the same. The only difference is the number of the valence shell being filled.

2.55 The element might be in Group 2A, all of which have two valence electrons. It might also be helium.

2.57 The properties are similar because all of them have the same outer-shell electron configuration. They are not identical because each has a different number of filled inner shells.

2.59 Ionization energy generally increases from left to right within a period in the Periodic Table and from bottom to top within column:
(a) K, Na, Li (b) C, N, Ne (c) C, O, F (d) Br, Cl, F

2.61 Following are the ground-state electron configurations of Mg atom, Mg^+, Mg^{2+}, and Mg^{3+}.

$$Mg \longrightarrow Mg^+ + e^- \quad IE = 738\ kJ/mol$$
Electron configuration: $1s^22s^22p^63s^2$ → $1s^22s^22p^63s^1$

$$Mg^+ \longrightarrow Mg^{2+} + e^- \quad IE = 1450\ kJ/mol$$
Electron configuration: $1s^22s^22p^63s^1$ → $1s^22s^22p^6$

$$Mg^{2+} \longrightarrow Mg^{3+} + e^- \quad IE = 7734\ kJ/mol$$
Electron configuration: $1s^22s^22p^6$ → $1s^22s^22p^5$

The first electron is removed from the 2s orbital. The removal of each subsequent electron requires more energy because, after the first electron is removed, each subsequent electron is removed from a positive ion, which strongly attracts the remaining electrons. The third ionization energy is especially large because the electron is removed from the filled second principal energy level, meaning that it is removed from an ion that has the same electron configuration as neon.

2.63 The most abundant elements by weight (a) in the Earth's crust are oxygen and silicon, and (b) in the human body they are oxygen and carbon.

2.65 Calcium is an essential element in human bones and teeth. Because strontium behaves chemically much like calcium, strontium-90 gets into our bones and teeth and gives off radioactivity for many years directly into our bodies.

2.67 Copper can be made harder by hammering it.

2.69 (a) Metals (b) Nonmetals (c) Metals
(d) Nonmetals (e) Metals (f) Metals

2.71 (a) Phosporous-32 has 15 protons, 15 electrons, and 32 - 15 = 17 neutrons.
(b) Molybdenum-98 has 42 protons, 42 electrons, and 98 - 42 = 56 neutrons.
(c) Calcium-44 has 20 protons, 20 electrons, and 44 - 20 = 24 neutrons.
(d) Hydrogen-3 has 1 proton, 1 electron, and 3 - 1 = 2 neutrons.
(e) Gadolinium-158 has 64 protons, 64 electrons, and 158 - 64 = 94 neutrons.
(f) Bismuth-212 has 83 protons, 83 electrons, and 212 - 83 = 129 neutrons.

2.73 Isotopes of elements from 37 to 53 contain more neutrons than protons.

2.75 Rounded to three significant figures, the atomic weight of naturally occurring boron is 10.8. The value given in the Periodic Table is 10.811.

$$\left(\frac{19.9}{100} \times 10.013 \text{ amu}\right) + \left(\frac{80.1}{100} \times 11.009 \text{ amu}\right) = 10.811 \text{ amu}$$

2.77 It would take 6.0×10^{21} protons to equal the mass of a grain of salt.

$$\frac{1.0 \times 10^{-2} \text{ g NaCl}}{1.67 \times 10^{-24} \text{ g/proton}} = 6.0 \times 10^{21} \text{ protons}$$

2.79 Assume the isotope mass is equal to the isotope mass number. By using these relationships, the solution is reached as follows:

$$\frac{\%\,^{85}\text{Rb}}{100} + \frac{\%\,^{87}\text{Rb}}{100} = 1 \quad \text{or} \quad \frac{\%\,^{85}\text{Rb}}{100} = 1 - \frac{\%\,^{87}\text{Rb}}{100}$$

$$\left(\frac{\%\,^{85}\text{Rb}}{100} \times 85\right) + \left(\frac{\%\,^{87}\text{Rb}}{100} \times 87\right) = 85.47$$

$$\left[\left(1 - \frac{\%\,^{87}\text{Rb}}{100}\right) \times 85\right] + \left(\frac{\%\,^{87}\text{Rb}}{100} \times 87\right) = 85.47$$

$$85 - \left(\frac{\%\,^{87}\text{Rb}}{100} \times 85\right) + \left(\frac{\%\,^{87}\text{Rb}}{100} \times 87\right) = 85.47$$

$$2 \times \frac{\%\,^{85}\text{Rb}}{100} = 85.47 - 85$$

$$^{87}\text{Rb} = 23.5\%$$

$$^{85}\text{Rb} = 100 - 23.5 = 76.5\%$$

2.81 Xenon (Xe) will have the highest ionization energy. Ionization energy increases from left to right going across the periodic table.

2.83 Element 118 will be in Group 8A. Expect it to be a gas that forms either no compounds or form very few compounds.

3.1 $^{139}_{53}I \rightarrow ^{0}_{-1}e + ^{139}_{54}Xe$

3.3 $^{74}_{33}As \rightarrow ^{0}_{+1}e + ^{74}_{32}Ge$

3.5 Barium-122 (10 g) has decayed through 5 half-lives, leaving 0.31 g:
10 g → 5.0 g → 2.5 g → 1.25 g → 0.625 g → 0.31 g

3.7 The intensity of any radiation decreases with the square of the distance: $\frac{I_1}{I_2} = \frac{d_2^2}{d_1^2}$

$$\frac{300 \text{ mCi}}{I_2} = \frac{(3.0 \text{ m})^2}{(0.01 \text{ m})^2}$$

$$I_2 = \frac{(300 \text{ mCi})(0.01 \cancel{\text{m}})^2}{(3.0 \cancel{\text{m}})^2} = 3.3 \times 10^{-3} \text{ mCi}$$

3.9 $$f = \frac{c}{\lambda} = \frac{3.0 \times 10^{10} \cancel{\text{cm}}/\text{s}}{5.8 \cancel{\text{cm}}} = 5.2 \times 10^{9}/\text{s}$$

3.11 $$f = \frac{c}{\lambda} = \frac{3.0 \times 10^{8} \cancel{\text{m}}/\text{s}}{650 \cancel{\text{nm}}}\left(\frac{10^{9} \cancel{\text{nm}}}{1 \cancel{\text{m}}}\right) = 4.6 \times 10^{14}/\text{s}$$

3.13 (a) $^{19}_{9}F$ (b) $^{32}_{15}P$ (c) $^{87}_{37}Rb$

3.15 For the lighter elements up to calcium, the stable isotopes have equal numbers of protons and neutrons. Boron-10 is the most stable isotope because of an equal number of protons and neutrons in the nucleus.

3.17 $^{151}_{62}Sm \rightarrow ^{0}_{-1}e + ^{151}_{63}Eu$

3.19 $^{51}_{24}Cr + ^{0}_{-1}e \rightarrow ^{51}_{23}V$

3.21 $^{248}_{96}Cm + ^{28}_{10}X \rightarrow ^{116}_{51}Sb + ^{160}_{55}Cs$
The bombarding nucleus is $^{28}_{10}Ne$.

3.23 (a) ${}^{10}_{4}\text{Be} \rightarrow {}^{0}_{-1}\text{e} + {}^{10}_{5}\text{B}$ Beta emission
(b) ${}^{151}_{63}\text{Eu}^* \rightarrow \gamma + {}^{151}_{63}\text{Eu}$ Gamma emission
(c) ${}^{195}_{81}\text{Tl} \rightarrow {}^{0}_{+1}\text{e} + {}^{195}_{80}\text{Hg}$ Positron emission
(d) ${}^{239}_{94}\text{Pu} \rightarrow {}^{4}_{2}\text{He} + {}^{235}_{92}\text{U}$ Alpha emission

3.25 Gamma emission does not result in transmutation.

3.27 ${}^{239}_{94}\text{Pu} + {}^{4}_{2}\text{He} \rightarrow {}^{240}_{95}\text{Am} + {}^{1}_{1}\text{p} + 2\,{}^{1}_{0}\text{n}$

3.29 After three half-lives: 1/2 x 1/2 x 1/2 = 1/8 or 12.5% of the original amount remains.

3.31 No, the conversion of Ra to Ra^{2+} involves the loss of valence electrons, which is not a nuclear process and does not involve a change in radioactivity.

3.33 50.0 mg → 25.0 mg → 12.5 mg → 6.25 mg →3.12 mg
Four half-lives in 60 minutes: half-life = 60 min/4 = 15 minutes

3.35 Gamma radiation has the greatest penetrating power; therefore it requires the largest amount of shielding.

3.37 It would be best to stand at least 30 meters away if you wish to be subjected to no more than 0.20mCi.

$$\frac{I_1}{I_2} = \frac{d_2^2}{d_1^2}$$

$$d_2 = \sqrt{\frac{d_1^2 I_1}{I_2}} = \sqrt{\frac{(1.0\text{ m})^2(175\text{ mCi})}{(0.20\text{ mCi})}} = 3.0 \times 10^1\text{ m}$$

3.39 (a) Amount of radiation absorbed by tissues
(b) Effective dose absorbed by humans
(c) Effective does delivered
(d) Intensity of radiation
(e) Amount of radiation absorbed by tissues
(f) Intensity of radiation
(g) Effective dose absorbed by humans

3.41 Alpha particles have so little penetrating power that they cannot penetrate the thick layer of skin on the hand. If they get into the lung, the thin membranes offer little resistance to the particles, which then damage the cells of the lung.

3.43 Alpha particles are the most damaging to tissue.

3.45 Iodine-131 is concentrated in the thyroid; therefore it would be expected to induce thyroid cancer.

3.47 (a) Cobalt-60 is used for (4) cancer therapy.
(b) Thallium-201 is used in (1) heart scans and exercise stress tests.
(c) Tritium is used for (2) measuring water content of the body.
(d) Mercury-197 is used for (3) kidney scans.

3.49 $^{248}_{96}\text{Cm} + ^{4}_{2}\text{He} \rightarrow 2\,^{1}_{0}\text{n} + ^{1}_{1}\text{H} + ^{249}_{97}\text{Bk}$

3.51 $^{208}_{82}\text{Pb} + ^{86}_{36}\text{Kr} \rightarrow 4\,^{1}_{0}\text{n} + ^{290}_{118}\textit{Unknown}$

3.53 The lifetime of a plant is short relative to the slow radioactive decay of carbon-14 to carbon-12; therefore the change in carbon-14 to carbon-12 ratio is negligible over the lifetime of a plant.

3.55 2006 – 1350 = 656 years (if the experiment was run in the year 2006)
656 years/5730 = 0.114 half-lives

3.57 Radon-222 produced polonium-218 by alpha emission:
$^{222}_{86}\text{Rn} \rightarrow ^{4}_{2}\text{He} + ^{218}_{84}\text{Po}$

3.59 The decay of a radioactive isotope is an exponential curve:

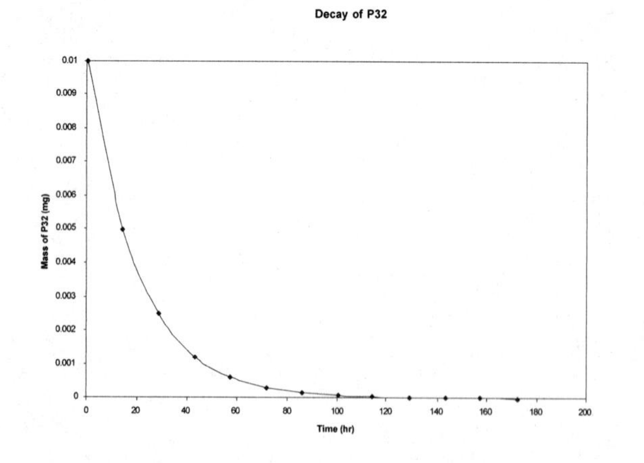

3.61 $^{19}_{10}\text{Ne} \rightarrow ^{0}_{+1}\text{e} + ^{19}_{9}\text{F}$

$^{20}_{11}\text{Na} \rightarrow ^{0}_{+1}\text{e} + ^{20}_{10}\text{Ne}$

3.63 Both the curie and the becquerel have units of disintegrations/second, a measurement of radiation intensity.

3.65 (a) $\frac{294 \text{ mrem/yr}}{359 \text{ mrem/yr}} \times 100 = 82\%$

(b) $\frac{39 \text{ mrem/yr}}{359 \text{ mrem/yr}} \times 100 = 11\%$

(c) $\frac{0.5 \text{ mrem/yr}}{359 \text{ mrem/yr}} \times 100 = 0.1\%$

3.67 X-rays will cause more ionization than radar because X-rays are higher energy.

3.69 1000/475 ~ 2 half-lives: 1/2 x 1/2 = 1/4 so 25% of the original americium will be around after 1000 years.

3.71 One sievert is equal to 100 rem. This is a sufficient dose to cause radiation sickness but not certain death.

3.73 (a) Radioactive elements are constantly decaying to other isotopes and elements mixed in with the original isotopes.
(b) Beta emissions result from the decay of a neutron in the nucleus to a proton (the increase in atomic number) and an electron (beta particle).

3.75 Oxygen-16 is stable because it has an equal number of protons and neutrons. The others are unstable because the numbers of protons and neutrons are unequal. In this case, the greater the difference in numbers of protons and neutrons, the faster the isotope decays.

3.77 $^{208}_{82}\text{Pb} + ^{64}_{28}\text{Ni} \rightarrow 6\,^{1}_{0}\text{n} + ^{266}_{110}\text{X}$

The new element is Darmstadtium: $^{266}_{110}\text{Ds}$

3.79 Lithium-7 and He-4 are produced when boron control rods absorb neutrons.

$^{10}_{5}\text{B} + ^{1}_{0}\text{n} \rightarrow ^{11}_{5}\text{B}$

$^{11}_{5}\text{B} \rightarrow ^{4}_{2}\text{He} + ^{7}_{3}\text{Li}$

<u>4.1</u> (a) Magnesium (Mg) atom with two valance electrons, loses both electrons to form a Mg^{2+} ion with a Neon (Ne) electron configuration.
(b) Sulfur (S) atom with six valance electrons gains two electrons to give a sulfide ion (S^{2-}) with an eight valance electron octet of an argon electron configuration.

<u>4.3</u> (a) KCl (b) CaF_2 (c) Fe_2O_3

<u>4.5</u> (a) $MgCl_2$ (b) Al_2O_3 (c) LiI

<u>4.7</u> (a) Potassium hydrogen phosphate (b) Aluminum sulfate
(c) Iron (II) carbonate <u>or</u> Ferrous carbonate

<u>4.9</u> (a) $\overset{\delta+}{C}-\overset{\delta-}{N}$ (b) $\overset{\delta+}{N}-\overset{\delta-}{O}$ (c) $\overset{\delta+}{C}-\overset{\delta-}{Cl}$

<u>4.11</u> Lewis structures:

(a) H–C–H (with H above and below C) — methane
(b) H₂C=CH₂ — ethene
(c) :O=C=O: — carbon dioxide
(d) H–C≡C–H — acetylene

<u>4.13</u> Contributing structures:

(a) H–C(=O:)–Ö:⁻ ⟷ H–C⁺(–:Ö:⁻)–Ö:⁻

(b) H–C(=O:)–Ö:⁻ ⟷ H–C(–:Ö:⁻)=Ö

(c) H₃C–C(=O:)–Ö–CH₃ ⟷ H₃C–C(–:Ö:⁻)=Ö⁺–CH₃

<u>4.15</u> Given are three-dimensional structures showing all unshared electron pairs.

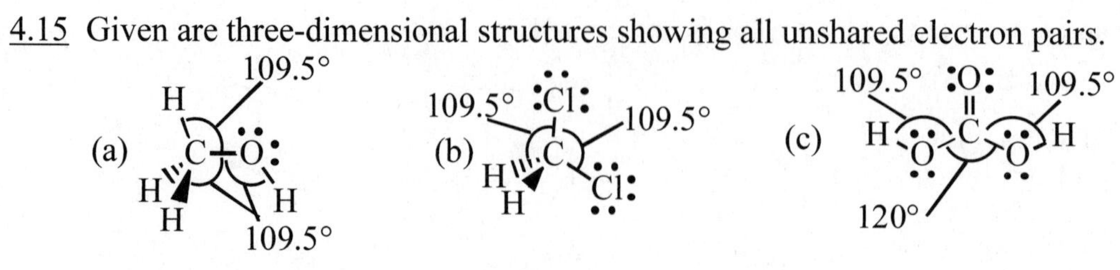

Chapter 4 Chemical Bonds

4.17 (a) By losing one electron, Li becomes Li^+, a helium electron configuration.
Li: $1s^2 2s^1 \rightarrow$ Li^+: $1s^2$ (filled *s* shell)
(b) By gaining one electron, Cl becomes Cl^-, an argon electron configuration.
Cl $1s^2 2s^2 2p^6 3s^2 3p^5 \rightarrow$ Cl^-: $1s^2 2s^2 2p^6 3s^2 3p^6$ (octet)
(c) By gaining three electrons, P becomes P^{3-}, an argon electron configuration.
P $1s^2 2s^2 2p^6 3s^2 3p^3 \rightarrow$ P^{3-}: $1s^2 2s^2 2p^6 3s^2 3p^6$ (octet)
(d) By losing three electrons, Al becomes Al^{3+}, a neon electron configuration.
Al: $1s^2 2s^2 2p^6 3s^2 3p^1 \rightarrow$ Al^{3+}: $1s^2 2s^2 2p^6$ (octet)
(e) By losing two electrons, Sr becomes Sr^{2+}, a krypton electron configuration.
Sr: $1s^2 2s^2 2p^6 3s^2 3p^6 4s^2 3d^{10} 4p^6 5s^2 \rightarrow$ Sr: $1s^2 2s^2 2p^6 3s^2 3p^6 4s^2 3d^{10} 4p^6$ (octet)
(f) By gaining two electrons, S becomes S^{2-}, an argon electron configuration.
S: $1s^2 2s^2 2p^6 3s^2 3p^4 \rightarrow$ S^{2-}: $1s^2 2s^2 2p^6 3s^2 3p^6$ (octet)
(g) By gaining four electrons, Si becomes Si^{4-}, an argon electron configuration.
Si: $1s^2 2s^2 2p^6 3s^2 3p^2 \rightarrow$ Si^{4-}: $1s^2 2s^2 2p^6 3s^2 3p^6$ (octet)
or by losing four electrons, Si becomes Si^{4+}, a neon electron configuration.
Si: $1s^2 2s^2 2p^6 3s^2 3p^2 \rightarrow$ Si^{4+}: $1s^2 2s^2 2p^6$ (octet)
(h) By gaining two electrons, O becomes O^{2-}, a neon electron configuration
O: $1s^2 2s^2 2p^4 \rightarrow$ O^{2-}: $1s^2 2s^2 2p^6$ (octet)

4.19 (a) H: $1s^1$ + 1 electron $\rightarrow$ H: $1s^2$ (filled 1*s* shell)
(b) Al: $1s^2 2s^2 2p^6 3s^2 3p^1 \rightarrow$ Al^{3+}: $1s^2 2s^2 2p^6$ (octet) + 3 electrons

4.21 Li^- is not stable because it has an unfilled 2nd shell.

4.23 Only (f) Cs+ will be stable because it has a Nobel gas valance shell. The other ions have a partially filled shell or a charge that is too high.

4.25 No. Copper is a transition metal so the octet rule does not apply. Transition metals can expand their octet into the 3*d*-orbitals.

4.27 Electronegativity increases going up a column of the Periodic Table because valence electrons are in shells closer to the electropositive nucleus. The decreasing distance of the valence electrons from the positively charged nucleus put the valence electrons under increasing pull.

4.29 (a) F occurs above Cl in the Periodic Table, therefore it is more electronegative.
(b) O occurs above S in the Periodic Table, therefore it is more electronegative.
(c) N occurs to the right of C in the Periodic Table, therefore it is more electronegative.
(d) F occurs to the right of C in the Periodic Table, therefore it is more electronegative.

4.31 The most polar bond occurs with the greatest electronegativity difference between atoms.
Bond polarity in decreasing order: C-O bond > C-N bond > C-C bond.

4.33 Use the difference in electronegativity to determine the character of the bond.
(a) C-Br (2.8-2.5 = 0.3); nonpolar covalent (b) S-Cl (3.0-2.5 = 0.5); polar covalent
(c) C-P (2.5-2.1 = 0.4); nonpolar covalent

4.35 (a) NaBr (b) Na_2O (c) $AlCl_3$ (d) $BaCl_2$ (e) MgO

4.37 Sodium chloride in the solid state has is a Na^+ ion surrounded by six Cl^- anions and each Cl^- anion surrounded by six Na+ ions.

4.39 (a) $Fe(OH)_3$ (b) $BaCl_2$ (c) $Ca_3(PO_4)_2$ (d) $NaMnO_4$

4.41 (a) The formula, $(NH_4)_2PO_4$, is incorrect. The correct formula is: $(NH_4)_3PO_4$.
(b) The formula, Ba_2CO_3, is incorrect. The correct formula is $BaCO_3$.
(c) The formula for aluminum sulfide, Al_2S_3 is correct.
(d) The formula for magnesium sulfide, MgS is correct.

4.43 KCl (potassium chloride) and $KHCO_3$ (potassium bicarbonate)

4.45 (a) SO_3^{2-} = sulfite (b) NO_3^- = nitrate (c) CO_3^{2-} = carbonate
(d) OH^- = hydroxide(e) HPO_4^{2-} = hydrogen phosphate

4.47 (a) Sodium fluoride (b) Magnesium sulfide (c) Aluminum oxide
(d) Barium chloride (e) Calcium hydrogen sulfite (Calcium bisulfite)
(f) Potassium iodide (g) Strontium phosphate
(h) Iron(II) hydroxide (Ferrous hydroxide) (i) Sodium dihydrogen phosphate
(j) Lead(II) acetate (Plumbous acetate) (k) Barium hydride
(l) Ammonium hydrogen phosphate

4.49 (a) NH_4HSO_3 (b) $Mg(C_2H_3O_2)_2$ (c) $Sr(H_2PO_4)_2$
(d) Ag_2CO_3 (e) $SrCl_2$ (f) $Ba(MnO_4)_2$

4.51 (a) A single bond results when one electron pair is shared between two atoms.
(b) A double bond results when two electron pairs are shared between two atoms.
(c) A triple bond results when three electron pairs are shared between two atoms.

4.53 Lewis structures for the following compounds:

(a) H–C(–H)(–H)–H (b) H–C≡C–H (c) H₂C=CH₂ (H and H on each C, C=C)

(d) :F̤–B(–F̤:)–F̤: (each F with three unshared pairs) (e) Ö=C(–H)–H (O with two unshared pairs) (f) :C̤l̤–C(–C̤l̤:)(–C̤l̤:)–C(–C̤l̤:)(–C̤l̤:)–C̤l̤: (each Cl with three unshared pairs)

4.55 Total number of valence electrons for each compound:

(a) NH_3 has 8 (b) C_3H_6 has 18 (c) $C_2H_4O_2$ has 24
(d) C_2H_6O has 20 (e) CCl_4 has 32 (f) HNO_2 has 18
(g) CCl_2F_2 has 32 (h) O_2 has 12

4.57 A bromine atom contains seven electrons in its valence shell. A bromine molecule contains two bromine atoms bonded by a single bond. A bromide ion is a bromine atom that has gained one electron in its valence shell; it has a complete octet and a charge of -1.

(a) :B̤r· (b) :B̤r–B̤r: (c) :B̤r:$^-$

4.59 Hydrogen has the electron configuration 1s^1. Hydrogen's valence shell has only a 1*s* orbital, which can hold only two electrons.

4.61 Nitrogen has five valence electrons. By sharing three more electrons with another atom(s), nitrogen can achieve the outer-shell electron configuration of neon, the noble gas nearest to it in atomic number. The three shared pairs of electrons may be in the form of three single bonds, one double bond and one single bond, or one triple bond. With any of these combinations, there is one unshared pair of electrons on nitrogen.

4.63 Oxygen has six valence electrons. By sharing two electrons with another atom(s), oxygen can achieve the outer-shell electron configuration of neon, the Noble gas nearest to it in atomic number. The two shared pairs of electrons may be in the form of a double bond or two single bonds. With either of these configurations, there are two unshared pairs of electrons on the oxygen.

4.65 O^{6+} has a charge too concentrated for a small ion.

4.67 (a) BF_3 has six valence electrons around the boron, thus does not obey octet rule.
(b) CF_2: does not obey the octet rule because carbon has 4 electrons around it.
(c) BeF_2: does not obey the octet rule because Be has 4 electrons around it.
(d) $H_2C{=}CH_2$ obeys the octet rule
(e) CH_3 does not obey the octet rule because carbon has 6 electrons around it.
(f) N_2 obeys the octet rule.
(g) NO does not obey the octet rule. Lewis structures drawn for the compound show either a nitrogen or an oxygen atom with 7 electrons.

4.69 Contributing structures for the bicarbonate ion:

H−O−C(=O)−O:⁻ ⟷ H−O−C(−O:⁻)=O

4.71 (a) There are 16 valence electrons present in N_2O.
(b) The two contributing structures can be represented as follows:

⁻N=N⁺=O ⟷ N≡N⁺−O:⁻

(c) Valid contributing structures must have the same number of valence electrons. The proposed structure has only 14 valance electrons where N_2O must have 16 valence electrons.

4.73 (a) H_2O has 8 valence electrons, and H_2O_2 has 14 valence electrons.

H−O−H H-O-O-H

(c) Predicted bond angles of 109.5° about each oxygen atom.

4.75 Shape of the each molecule and approximate bond angles about its central atom:
(a) Tetrahedral, 109.5° (b) Pyramidal, 109.5° (c) Tetrahedral, 109.5°
(d) Bent, 120° (e) Trigonal planar, 120° (f) Tetrahedral, 109.5°
(g) Pyramidal, 109.5° (h) Pyramidal, 109.5°

4.77 The differences are in their shapes. Because CO_2 is a linear molecule, it is nonpolar. SO_2 is a bent molecule, therefore it is polar.

(a) O=C=O (b) O=S=O

4.79 Yes, it is possible to have a molecule to have polar bonds, yet no dipole. This occurs when the individual polar bonds act in equal but opposite directions, as in CO_2.

O=C=O   No net dipole moment

4.81 Use differences in electronegativity to predict the polarity of the bond.

(a) Nonpolar covalent	(b) Polar covalent	(c) Polar covalent	(d) Ionic
(e) Polar covalent	(f) Polar covalent	(g) Nonpolar covalent	(h) Ionic

4.83 Calcium dihydrogen phosphate, calcium phosphate, and calcium carbonate.

4.85 Barium sulfate is used to visualize the gastrointestinal tract by X-ray examination.

4.87 Calcium (Ca^{2+}) is the main metal ion present in bone and tooth enamel

4.89 Argon already has an octet with eight valence electrons in its outer shell, therefore (a) it does not donate or accept electrons to form ions and (b) it doesn't need to form covalent bonds by sharing electrons.

4.91 The two possibilities for a structure are square pyramidal and trigonal bipyramidal.

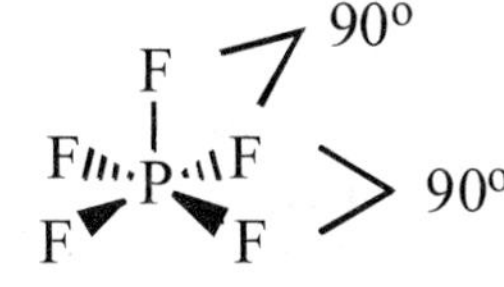
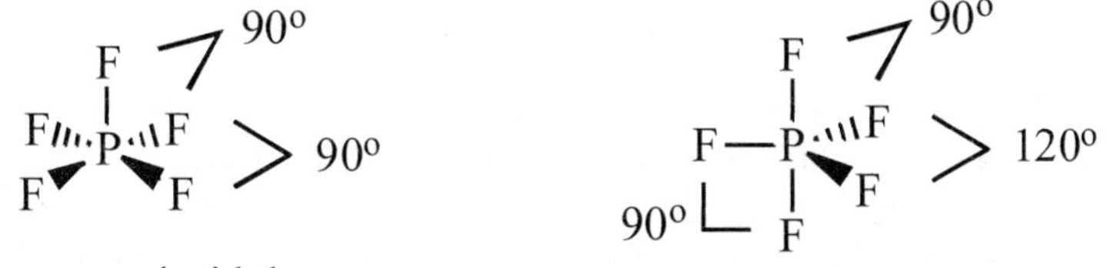

Square pyrimidal

Trigonal bipyrimidal

The square pyramidal structure contains eight F-F 90° bonding electron pair repulsions. These repulsions are minimized in a trigonal bipyramidal structure, where there are only six F-F 90° bonding electron pair repulsions. The VSEPR theory would then predict molecules like PF_5 would adopt the least strained configuration, a trigonal bipyramidal structure.

4.93 (a) ClO_2 has 19 valence electrons. In the Lewis structure, either the chlorine atom or one of the oxygen atoms must have only seven valence electrons. In the following Lewis structure, the odd electron is placed on the chlorine, which is the least electronegative element of the two atoms.

(b) Lewis structure of ClO_2:

O=Cl=O

4.95 Zinc oxide, ZnO, is present in sun blocking agents and helps reflect sunlight away from the skin.

4.97 Lead(IV) oxide, PbO_2, and lead(IV) carbonate, $Pb(CO_3)_2$, were used as pigments in paint.

4.99 Fe(II) is utilized in over-the-counter iron supplements.

4.101 (a) $CaSO_3$ (b) $Ca(HSO_3)_2$ (c) $Ca(OH)_2$ (d) $CaHPO_4$

4.103 Perchloroethylene does possess four polar covalent C-Cl bonds, but is not a polar compound. The molecule lacks a dipole because the polar covalent C-Cl bonds act in equal, but opposite directions.

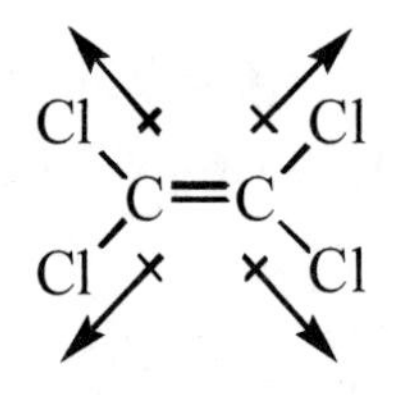

4.105 (a) Following is a Lewis structure for tetrafluoroethylene.
(b) All bond angles are predicted to be 120°.
(c) Tetrafluoroethylene has four polar covalent C-F bonds, but is a nonpolar molecule, because like in problem 4.103, the polar covalent C-F bonds act in equal, but opposite directions.

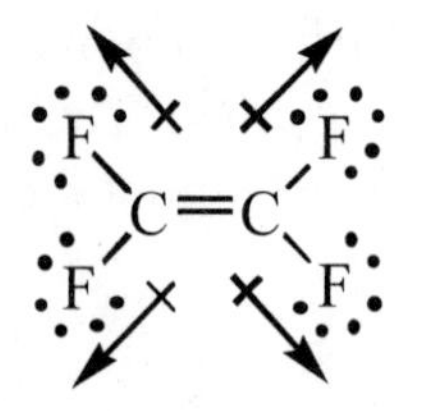

5.1 (a) Ibuprofen, $C_{13}H_{18}O_2$
C 13 x 12.0 amu = 156 amu
H 18 x 1.00 amu = 18.0 amu
O 2 x 16.0 amu = 32.0 amu
$C_{13}H_{18}O_2$ = 206 amu

(b) Barium Phosphate, $Ba_3(PO_4)_2$
Ba 3 x 137 amu = 411 amu
P 2 x 31.0 amu = 62.0 amu
O 8 x 16.0 amu = 128 amu
$Ba_3(PO_4)_2$ = 601 amu

5.3 $$2.84 \text{ mol } Na_2S \left(\frac{87.1 \text{ g } Na_2S}{1 \text{ mol } Na_2S}\right) = 222 \text{ g } Na_2S$$

5.5 Moles of Cu(I) ions:

$$0.062 \text{ g } CuNO_3 \left(\frac{1 \text{ mol } CuNO_3}{125.5 \text{ g } CuNO_3}\right)\left(\frac{1 \text{ mol } Cu^+ \text{ ions}}{1 \text{ mol } CuNO_3}\right) = 4.9 \times 10^{-4} \text{ mol } Cu^+$$

5.7 Balance: $CO_2(g) + H_2O(l) \longrightarrow C_6H_{12}O_6(aq) + O_2(g)$
Step 1: Balance carbons with a coefficient of 6 in front of CO_2

$$6CO_2(g) + H_2O(l) \longrightarrow C_6H_{12}O_6(aq) + O_2(g)$$

Step 2: Balance hydrogen with a coefficient of 6 in front of H_2O

$$6CO_2(g) + 6H_2O(l) \longrightarrow C_6H_{12}O_6(aq) + O_2(g)$$

Step 3: Last step, balance oxygen with a coefficient of 6 in front of O_2

Balanced equation: $6CO_2(g) + 6H_2O(l) \xrightarrow{\text{photosynthesis}} C_6H_{12}O_6(aq) + 6O_2(g)$

5.9 Balance the equation: $K_2C_2O_4(aq) + Ca_3(AsO_4)_2(s) \longrightarrow K_3AsO_4(aq) + CaC_2O_4(s)$
Step 1: First balance the most complicated AsO4 with a 2 in front of K_3AsO_4

$$K_2C_2O_4(aq) + Ca_3(AsO_4)_2(s) \longrightarrow 2K_3AsO_4(aq) + CaC_2O_4(s)$$

Step 2: Next, balance the potassium by placing a 3 in front of $K_2C_2O_4$

$$3K_2C_2O_4(aq) + Ca_3(AsO_4)_2(s) \longrightarrow 2K_3AsO_4(aq) + CaC_2O_4(s)$$

Step 3: Now balance C_2O_4 and Ca with a coefficient of 3 in front of CaC_2O_4
Balanced equation:

$$3K_2C_2O_4(aq) + Ca_3(AsO_4)_2(s) \longrightarrow 2K_3AsO_4(aq) + 3CaC_2O_4(s)$$

5.11 Using the balanced equation: $CH_3OH(g) + CO(g) \longrightarrow CH_3COOH(l)$,
the molar ratio of CO required to produce CH_3COOH is 1:1; therefore 16.6 moles of CO is required to produce 16.6 moles of CH_3COOH.

5.13 6.0 g carbon = 0.50 mol of carbon   2.1 g H_2 = 1.1 mol H_2

$$\text{Mass of } H_2 \text{ required} = 6.0\ \cancel{g\,C}\left(\frac{1\ \cancel{mol\,C}}{12.0\ \cancel{g\,C}}\right)\left(\frac{2\ \cancel{mol\,H_2}}{1\ \cancel{mol\,C}}\right)\left(\frac{2.0\ g}{1\ \cancel{mol\,H_2}}\right) = 2.0\ g\ H_2$$

$$\text{Mass of C required} = 2.1\ \cancel{g\,H_2}\left(\frac{1\ \cancel{mol\,H_2}}{2.0\ \cancel{g\,H_2}}\right)\left(\frac{1\ \cancel{mol\,C}}{2\ \cancel{mol\,H_2}}\right)\left(\frac{12.0\ g\ C}{1\ \cancel{mol\,C}}\right) = 6.2\ g\ C$$

(a) H_2 is in excess and C is the limiting reagent.
(b) 8.0 grams of CH_4 are produced.

$$6.0\ \cancel{g\,C}\left(\frac{1\ \cancel{mol\,C}}{12.0\ \cancel{g\,C}}\right)\left(\frac{1\ \cancel{mol\,CH_4}}{1\ \cancel{mol\,C}}\right)\left(\frac{16.0\ g\ CH_4}{1\ \cancel{mol\,CH_4}}\right) = 8.0\ g\ CH_4 \text{ produced}$$

5.15 Overall chemical reaction: $CuCl_2(aq) + K_2S(aq) \rightarrow CuS(s) + 2KCl(aq)$
Step 1: Write an equation involving all of the chemical species participating in the chemical reaction.
$Cu^{2+}(aq) + 2Cl^-(aq) + 2K^+(aq) + S^{2-}(aq) \rightarrow CuS(s) + 2K^+(aq) + 2Cl^-(aq)$

Step 2: Cross out the aqueous ions that appear on both sides of the equation.
$Cu^{2+}(aq) + \cancel{2Cl^-(aq)} + \cancel{2K^+(aq)} + S^{2-}(aq) \rightarrow CuS(s) + \cancel{2K^+(aq)} + \cancel{2Cl^-(aq)}$
Net ionic equation: $Cu^{2+}(aq) + S^{2-}(aq) \rightarrow CuS(s)$
The net ionic equation shows the chemical species that actually undergo a chemical change. The ions that appear on both sides of the equation do not change, therefore are considered spectator ions.

5.17 (a) KCl = 74.6 amu   (b) Na_3PO_4 = 164.0 amu   (c) $Fe(OH)_2$ = 89.9 amu

5.19 (a) $C_{12}H_{22}O_{11}$ = 342.3 amu   (b) $C_2H_5NO_2$ = 75.1 amu   (c) $C_{14}H_9Cl_5$ = 354.5 amu

5.21 (a) $1.77\ \cancel{mol\,NO_2}\left(\frac{46.0\ g\ NO_2}{1\ \cancel{mol\,NO_2}}\right) = 81.4\ g\ NO_2$

(b) $0.84 \text{ mol } C_3H_8O \left(\frac{60.1 \text{ g } C_3H_8O}{1 \text{ mol } C_3H_8O} \right) = 50 \text{ g } C_3H_8O$

(c) $3.69 \text{ mol } UF_6 \left(\frac{352.0 \text{ g } UF_6}{1 \text{ mol } UF_6} \right) = 1.30 \times 10^3 \text{ g } UF_6$

(d) $0.348 \text{ mol } C_6H_{12}O_6 \left(\frac{180.2 \text{ g } C_6H_{12}O_6}{1 \text{ mol } C_6H_{12}O_6} \right) = 62.7 \text{ g } C_6H_{12}O_6$

(e) $4.9 \times 10^{-2} \text{ mol } C_6H_8O_6 \left(\frac{176.1 \text{ g } C_6H_8O_6}{1 \text{ mol } C_6H_8O_6} \right) = 8.6 \text{ g } C_6H_8O_6$

5.23 (a) $6.56 \text{ mol } Na_2S \left(\frac{1 \text{ mol } S^{2-} \text{ ions}}{1 \text{ mol } Na_2S} \right) = 6.56 \text{ mol } S^{2-} \text{ ions}$

(b) $8.320 \text{ mol } Mg_3(PO_4)_2 \left(\frac{3 \text{ mol } Mg^{2+} \text{ ions}}{1 \text{ mol } Mg_3(PO_4)_2} \right) = 24.96 \text{ mol } Mg^{2+} \text{ ions}$

(c) $0.43 \text{ mol } Ca(CH_3COO)_2 \left(\frac{2 \text{ mol } CH_3COO^- \text{ ions}}{1 \text{ mol } Ca(CH_3COO)_2} \right) = 0.86 \text{ mol } CH_3COO^- \text{ ions}$

5.25 The same; that is, just 2:1.

5.27 (a) $2.9 \text{ mol TNT} \left(\frac{6.02 \times 10^{23} \text{ molecules TNT}}{1 \text{ mol TNT}} \right) = 1.7 \times 10^{24} \text{ molecules of TNT}$

(b) $5.00 \times 10^{-2} \text{ g } H_2O \left(\frac{1 \text{ mol } H_2O}{18.0 \text{ g } H_2O} \right) = 2.78 \times 10^{-3} \text{ mol } H_2O$

$2.78 \times 10^{-3} \text{ mol } H_2O \left(\frac{6.02 \times 10^{23} \text{ molec. } H_2O}{1 \text{ mol } H_2O} \right) = 1.67 \times 10^{21} \text{ molecules } H_2O$

(c) $3.1 \times 10^{-1}\ \text{g}\ C_9H_8O_4 \left(\frac{1\ \text{mol}\ C_9H_8O_4}{180\ \text{g}\ C_9H_8O_4}\right) = 1.7 \times 10^{-3}\ \text{mol}\ C_9H_8O_4$

$1.7 \times 10^{-3}\ \text{mol}\ C_9H_8O_4 \left(\frac{6.02 \times 10^{23}\ \text{molec}}{1\ \text{mol}\ C_9H_8O_4}\right) = 1.0 \times 10^{21}\ \text{molec}\ C_9H_8O_4$

5.29 $\left(\frac{68{,}000\ \text{amu}}{1\ \text{mol hemoglobin}}\right)\left(\frac{1\ \text{mol hemoglobin}}{6.02 \times 10^{23}\ \text{molecules}}\right) = 1.1 \times 10^{-19}$ amu/molecule

5.31 The following are balanced equations:

(a) $H_2 + I_2 \rightarrow 2HI$

(b) $4Al + 3O_2 \rightarrow 2Al_2O_3$

(c) $2Na + Cl_2 \rightarrow 2NaCl$

(d) $2Al + 6HBr \rightarrow 2AlBr_3 + 3H_2$

(e) $4P + 5O_2 \rightarrow 2P_2O_5$

5.33 $CaCO_3(s) \xrightarrow{\text{heat}} CaO(s) + CO_2(g)$

5.35 $4Fe(s) + 3O_2(g) \rightarrow 2Fe_2O_3(s)$

5.37 $(NH_4)_2CO_3(s) \rightarrow 2NH_3(g) + CO_2(g) + H_2O(l)$

5.39 $2Al(s) + 3HCl(aq) \rightarrow 2AlCl_3(aq) + 3H_2(g)$

5.41 Using the balanced equation: $2N_2(g) + 3O_2(g) \rightarrow 2N_2O_3(g)$

(a) $1\ \text{mol}\ O_2 \left(\frac{2\ \text{mol}\ N_2}{3\ \text{mol}\ O_2}\right) = 0.67\ \text{mol}\ N_2$ required

(b) $1\ \text{mol}\ O_2 \left(\frac{2\ N_2O_3}{3\ \text{mol}\ O_2}\right) = 0.67\ \text{mol}\ N_2O_3$ produced

(c) $8\ \text{mol}\ N_2O_3 \left(\frac{3\ \text{mol}\ O_2}{2\ \text{mol}\ N_2O_3}\right) = 12\ \text{mol}\ O_2$ required

5.43 Using the balanced equation: $CH_4(g) + 3Cl_2(l) \rightarrow CHCl_3(g) + 3HCl(g)$

$$1.50 \text{ mol } CHCl_3 \left(\frac{3 \text{ mol } Cl_2}{1 \text{ mol } CHCl_3}\right)\left(\frac{70.9 \text{ g } Cl_2}{1 \text{ mol } Cl_2}\right) = 319 \text{ g } Cl_2 \text{ needed}$$

5.45 (a) Balanced equation: $2NaClO_2(aq) + Cl_2(g) \rightarrow 2ClO_2(g) + 2NaCl(aq)$

(b) 4.1 kg ClO_2

$$5.50 \text{ kg } NaClO_2 \left(\frac{1000 \text{ g } NaClO_2}{1 \text{ kg } NaClO_2}\right)\left(\frac{1 \text{ mol } NaClO_2}{90.4 \text{ g } NaClO_2}\right) = 60.8 \text{ mol } NaClO_2$$

$$60.8 \text{ mol } NaClO_2 \left(\frac{2 \text{ mol } ClO_2}{2 \text{ mol } NaClO_2}\right)\left(\frac{67.5 \text{ g } ClO_2}{1 \text{ mol } ClO_2}\right)\left(\frac{1 \text{ kg}}{1000 \text{ g}}\right) = 4.10 \text{ kg } ClO_2$$

5.47 Using the balanced equation: $6CO_2(g) + 6H_2O(l) \rightarrow C_6H_{12}O_6(aq) + 6O_2(g)$

$$5.1 \text{ g Glucose} \left(\frac{1 \text{ mol Glucose}}{180 \text{ g Glucose}}\right)\left(\frac{6 \text{ mol } CO_2}{1 \text{ mol Glucose}}\right)\left(\frac{44.0 \text{ g } CO_2}{1 \text{ mol } CO_2}\right) = 7.5 \text{ g } CO_2$$

5.49 Using the balanced equation in problem #5.48

$$0.58 \text{ g } Fe_2O_3 \left(\frac{1 \text{ mol } Fe_2O_3}{159.7 \text{ g } Fe_2O_3}\right)\left(\frac{6 \text{ mol C}}{2 \text{ mol } Fe_2O_3}\right)\left(\frac{12.0 \text{ g C}}{1 \text{ mol C}}\right) = 0.13 \text{ g C needed}$$

5.51 $$25.0 \text{ g Asp actual yield} \left(\frac{100 \text{ g Asp}}{75 \text{ g Asp actual yield}}\right)\left(\frac{1 \text{ mol Asp}}{180 \text{ g Asp}}\right) = 0.185 \text{ mol Asp}$$

$$0.185 \text{ mol Asp} \left(\frac{1 \text{ mol SA}}{1 \text{ mol Asp}}\right)\left(\frac{138 \text{g SA}}{1 \text{ mol SA}}\right) = 25.6 \text{ g SA}$$

You will need to use 25.6 g salicylic acid to get 25.0 g of aspirin after a 75% yield.

5.53 Using the balanced equation: $CH_3CH_3(g) + Cl_2(g) \rightarrow CH_3CH_2Cl(l) + HCl(g)$
Theoretical yield of ethyl chloride:

$$5.6\ \text{g Ethane}\left(\frac{1\ \text{mol Ethane}}{30.1\ \text{g Ethane}}\right)\left(\frac{1\ \text{mol}\ CH_3CH_2Cl}{1\ \text{mol Ethane}}\right)\left(\frac{64.5\ \text{g}\ CH_3CH_2Cl}{1\ \text{mol}\ CH_3CH_2Cl}\right) = 12\ \text{g}$$

Actual yield of CH_3CH_2Cl = 8.2 g $\qquad \%\ \text{yield} = \dfrac{\text{Actual yield}}{\text{Theoretical yield}} \times 100$

$$\%\ \text{Yield of}\ CH_3CH_2Cl = \frac{8.2\ \text{g}\ CH_3CH_2Cl}{12\ \text{g}\ CH_3CH_2Cl} \times 100 = 68\%$$

5.55 (a) Spectator ion: an ion that does not take part in a chemical reaction
(b) Net ionic equation: a balanced equation showing only the ions that react
(c) Aqueous solution: a solution using water as a solvent

5.57 (a) The spectator ions are Na^+ and Cl^-.
(b) $2Na^+(aq) + CO_3^{2-}(aq) + Sr^{2+}(aq) + 2Cl^-(aq) \rightarrow SrCO_3(s) + 2Na^+(aq) + 2Cl^-(aq)$
Balanced net ionic equation: $CO_3^{2-}(aq) + Sr^{2+}(aq) \rightarrow SrCO_3(s)$

5.59 $Pb^{2+}(aq) + 2NO_3^-(aq) + 2NH_4^+(aq) + 2Cl^-(aq) \rightarrow$

$PbCl_2(s) + 2NO_3^-(aq) + 2NH_4^+(aq)$

Balanced net ionic equation: $Pb^{2+}(aq) + 2Cl^-(aq) \rightarrow PbCl_2(s)$

5.61 $2Na^+(aq) + 2OH^-(aq) + 2NH_4^+(aq) + CO_3^{2-}(aq) \rightarrow$

$2NH_3(g) + 2H_2O(l) + 2Na^+(aq) + CO_3^{2-}(aq)$

Balanced net ionic equation: $NH_4^+(aq) + OH^-(aq) \rightarrow NH_3(g) + H_2O(l)$

5.63 (a) $MgCl_2$ (soluble): most compounds containing Cl^- are soluble
(b) $CaCO_3$ (insoluble): most compounds containing CO_3^{2-} are insoluble
(c) Na_2SO_4 (soluble): all compounds containing Na^+ are soluble
(d) NH_4NO_3 (soluble): all compounds containing NO_3^- and NH_4^+ are soluble
(e) $Pb(OH)_2$ (insoluble): most compounds containing OH^- are insoluble

5.65 No, one species gains electrons and the other loses electrons. Electrons cannot be destroyed, but transferred from one chemical species to another.

5.67 (a) C_7H_{12} is oxidized (the carbons gain oxygen going to CO_2) and O_2 is reduced.
(b) O_2 is the oxidizing agent and C_7H_{12} is the reducing agent.

5.69 An exothermic chemical reaction or process releases heat as a product.
An endothermic chemical reaction or process absorbs heat as a reactant.

5.71 19.6 kcal are given off.

5.73 $$15.0\ \text{g Glucose}\left(\frac{1\ \text{mol Glucose}}{180\ \text{g Glucose}}\right)\left(\frac{670\ \text{kcal}}{1\ \text{mol Glucose}}\right) = 55.8\ \text{kcal of heat evolved}$$

5.75 (a) The synthesis of starch is endothermic.
(b) 26.4 kcal

$$6.32\ \text{g starch}\left(\frac{10^{-3}\ \text{kg starch}}{1\ \text{g starch}}\right)\left(\frac{4178\ \text{kcal}}{1.00\ \text{kg starch}}\right) = 26.4\ \text{kcal heat required}$$

5.77 Fluoride reacts with the $Ca_{10}(PO_4)_6(OH)_2$ in enamel, by exchanging the OH- ions with F^-, forming a less soluble $Ca_{10}(PO_4)_6F_2$ under the acidic conditions found in the mouth.

5.79 Oxidation occurs at the anode, where $Fe^0 \longrightarrow Fe^{2+}$
Reduction occurs at the cathode, where $Zn^{2+} \longrightarrow Zn^0$

5.81 $N_2O_5(g) + 2H_2O(l) \rightarrow 2HNO_3(aq)$

5.83 (a) Fe_2O_3 loses oxygen; it is reduced. CO gains oxygen: it is oxidized.

(b) $$38.4\ \text{mol Fe}\left(\frac{1\ \text{mol Fe}_2\text{O}_3}{2\ \text{mol Fe}}\right) = 19.2\ \text{mol Fe}_2\text{O}_3\ \text{needed}$$

(c) $$38.4\ \text{mol Fe}\left(\frac{3\ \text{mol CO}}{2\ \text{mol Fe}}\right)\left(\frac{28.01\ \text{g CO}}{1\ \text{mol CO}}\right) = 1.61 \times 10^3\ \text{g CO required}$$

5.85 The spectator ions are Na^+ and NO_3^-.

$$6Na^+(aq) + 2PO_4^{3-}(aq) + 3Cd^{2+}(aq) + 6NO_3^-(aq) \rightarrow$$
$$Cd_3(PO_4)_2(s) + 6NO_3^-(aq) + 6Na^+(aq)$$

Balanced net ionic equation: $3Cd^{2+}(aq) + 2PO_4^{3-}(aq) \rightarrow Cd_3(PO_4)_2(s)$

5.87 $$MW_{chlorophyll} = \frac{24.305 \text{ g Mg /mol}}{0.0272 \text{ g Mg /1 g chlorophyll}} = 893 \text{ amu}$$

5.89 8.00 g $Pb(NO_3)_2$ added to 2.67 g $AlCl_3$ yielded 5.55g $PbCl_2$

Mass of aluminum chloride required based on 8.00 g $Pb(NO_3)_2$:

$$8.00 \text{ g } Pb(NO_3)_2 \left(\frac{1 \text{ mol } Pb(NO_3)_2}{331.2 \text{ g } Pb(NO_3)_2}\right)\left(\frac{2 \text{ mol } AlCl_3}{3 \text{ mol } Pb(NO_3)_2}\right) = 1.61 \times 10^{-2} \text{ mol } AlCl_3$$

$$1.61 \times 10^{-2} \text{ mol } AlCl_3 \left(\frac{133.3 \text{ g } AlCl_3}{1 \text{ mol } AlCl_3}\right) = 2.15 \text{ g } AlCl_3 \text{ needed}$$

Mass of lead(II) nitrate required based on 2.67 g $AlCl_3$:

$$2.67 \text{ g } AlCl_3 \left(\frac{1 \text{ mol } AlCl_3}{133.3 \text{ g } AlCl_3}\right)\left(\frac{3 \text{ mol } Pb(NO_3)_2}{2 \text{ mol } AlCl_3}\right) = 3.00 \times 10^{-2} \text{ mol of } Pb(NO_3)_2$$

$$3.00 \times 10^{-2} \text{ mol } Pb(NO_3)_2 \left(\frac{331.2 \text{ g } Pb(NO_3)_2}{1 \text{ mol } Pb(NO_3)_2}\right) = 9.94 \text{ g } Pb(NO_3)_2 \text{ needed}$$

(a) $Pb(NO_3)_2$ is the limiting reagent.

(b) Actual yield of $PbCl_2$:

$$8.00 \text{ g } Pb(NO_3)_2 \left(\frac{1 \text{ mol } Pb(NO_3)_2}{331.2 \text{ g } Pb(NO_3)_2}\right)\left(\frac{3 \text{ mol } PbCl_2}{3 \text{ mol } Pb(NO_3)_2}\right) = 2.42 \times 10^{-2} \text{ mol } PbCl_2$$

$$2.42 \times 10^{-2} \text{ mol } PbCl_2 \left(\frac{278.1 \text{ g } PbCl_2}{1 \text{ mol } PbCl_2}\right) = 6.73 \text{ g } PbCl_2$$

$$\% \text{ Yield} = \frac{\text{Actual yield}}{\text{Theoretical yield}} = \frac{5.55\text{g } PbCl_2}{6.73 \text{ g } PbCl_2} \times 100 = 82.5\% \text{ } PbCl_2$$

5.91 (a) $C_5H_{12}(g) + 8O_2(g) \rightarrow 5CO_2(g) + 6H_2O(g)$

(b) Pentane is oxidized and oxygen is reduced.

(c) Oxygen is the oxidizing agent and pentane is the reducing agent.

5.93 (a) The balanced combustion reactions are listed below:

$$CH_4(g) + 2O_2(g) \rightarrow CO_2(g) + 2H_2O(g) \quad + 213 \text{ kcal/mol}$$

$$C_3H_8(g) + 5O_2(g) \rightarrow 3CO_2(g) + 4H_2O(g) \quad + 530 \text{ kcal/mol}$$

(b) Propane releases more energy per mole (530 kcal/mol).

(c) Methane releases more energy per gram than propane.

$$\text{Methane heat of combustion (kcal/g)} = \frac{213 \text{ kcal/}\cancel{\text{mol}}}{16.0 \text{ g/}\cancel{\text{mol}}} = 13.3 \text{ kcal/g}$$

$$\text{Propane heat of combustion (kcal/g)} = \frac{530 \text{ kcal/}\cancel{\text{mol}}}{44.1 \text{ g/}\cancel{\text{mol}}} = 12.0 \text{ kcal/g}$$

Chapter 6 Gases, Liquids, and Solids

6.1 $$P_2 = \frac{P_1V_1}{V_2} = \frac{(0.70\ \text{atm})(3.8\ \cancel{L})}{6.5\ \cancel{L}} = 0.41\ \text{atm}$$

6.3 $$P_2 = \frac{P_1V_1T_2}{T_1V_2} = \frac{(0.92\ \text{atm})(20.5\ \cancel{L})(285\ \cancel{K})}{(296\ \cancel{K})(340.6\ \cancel{L})} = 0.053\ \text{atm}$$

$$P = \frac{nRT}{V} = \frac{(2.00\ \cancel{\text{mol}})(0.0821\ \cancel{L}\bullet\text{atm}\bullet\cancel{\text{mol}^{-1}\bullet K^{-1}})(295\ \cancel{K})}{10\ \cancel{L}} = 4.84\ \text{atm}$$

6.5 Ideal Gas Law: $PV = nRT$

$$n = \frac{PV}{RT} = \frac{(1.05\ \cancel{\text{atm}})(10.0\ \cancel{L})}{(0.0821\ \cancel{L\bullet\text{atm}}\bullet\text{mol}^{-1}\bullet\cancel{K^{-1}})(303\ \cancel{K})} = 0.422\ \text{mol Ne}$$

6.7 Dalton's Law of Partial Pressures:

Total pressure $(P_T) = P_{N_2} + P_{H_2O}$

$P_{H_2O} = P_T - P_{N_2} = 2.015\ \text{atm} - 1.908\ \text{atm} = 0.107$ atm of H_2O vapor

6.9 Heat of vaporization of water = 540 cal/g

$$45.0\ \cancel{\text{kcal}}\left(\frac{1000\ \cancel{\text{cal}}}{1\ \cancel{\text{kcal}}}\right)\left(\frac{1\ \text{g}\ H_2O}{540\ \cancel{\text{cal}}}\right) = 83.3\ \text{g}\ H_2O\ \text{vaporized}$$

6.11 According to the phase diagram of water (figure 5.18), the vapor will undergo reverse sublimation and form a solid.

6.13 Boyle's Law:

At constant temperature, $\left(\frac{P_1V_1}{\cancel{T_1}}\right) = \left(\frac{P_2V_2}{\cancel{T_2}}\right)$ reduces to $P_1V_1 = P_2V_2$

$$P_1 = \frac{P_2V_2}{V_1} = \frac{(12.2\ \text{atm})(2.5\ \cancel{L})}{20\ \cancel{L}} = 1.5\ \text{atm}\ CH_4$$

6.15 Gay-Lussac's Law: The tire is at constant volume.

At constant volume, $\left(\frac{P_1 \not{V_1}}{T_1}\right) = \left(\frac{P_2 \not{V_2}}{T_2}\right)$ reduces to $\frac{P_1}{T_1} = \frac{P_2}{T_2}$

$$P_2 = \frac{P_1 T_2}{T_1} = \frac{(2.30 \text{ atm})(320 \not{K})}{293 \not{K}} = 2.51 \text{ atm of air in the tire}$$

6.17 Charles's Law:

At constant presure, $\left(\frac{\not{P_1} V_1}{T_1}\right) = \left(\frac{\not{P_2} V_2}{T_2}\right)$ reduces to $\frac{V_1}{T_1} = \frac{V_2}{T_2}$

$$V_2 = \frac{V_1 T_2}{T_1} = \frac{(4.17 \text{ L})(448 \not{K})}{998 \not{K}} = 1.87 \text{ L of ethane gas upon cooling}$$

6.19 Gay-Lussac's Law:

At constant volume, $\left(\frac{P_1 \not{V_1}}{T_1}\right) = \left(\frac{P_2 \not{V_2}}{T_2}\right)$ reduces to $\frac{P_1}{T_1} = \frac{P_2}{T_2}$

$$T_2 = \frac{P_2 T_1}{P_1} = \frac{(375 \text{ mm Hg})(898 \text{ K})}{450 \text{ mm Hg}} = 748 \text{ K} \ \ (475^{\circ}\text{C})$$

6.21 Gay-Lussac's Law:

At constant volume, $\left(\frac{P_1 \not{V_1}}{T_1}\right) = \left(\frac{P_2 \not{V_2}}{T_2}\right)$ reduces to $\frac{P_1}{T_1} = \frac{P_2}{T_2}$

$$P_2 = \frac{P_1 T_2}{T_1} = \frac{(1.00 \text{ atm})(438 \not{K})}{373 \not{K}} = 1.17 \text{ atm}$$

6.23 Complete this table: Use the $\frac{P_1 V_1}{T_1} = \frac{P_2 V_2}{T_2}$ equation.

V_1	T_1	P_1	V_2	T_2	P_2
546 L	43°C	6.5 atm	**1198 L**	65°C	1.9 atm
43 mL	-56°C	865 torr	**48 mL**	43°C	1.5 atm
4.2 L	234 K	0.87 atm	3.2 L	29°C	**1.5 atm**
1.3 L	25°C	740 mm Hg	**1.2 L**	0°C	1.0 atm

6.25 Charles's Law: atmospheric pressure acting on balloon is constant

$$V_2 = \frac{V_1T_2}{T_1} = \frac{(1.2\ \text{L})(77\ \text{K})}{298\ \text{K}} = 0.31\ \text{L balloon's final volume}$$

6.27 $$P_2 = \frac{P_1V_1T_2}{T_1V_2} = \frac{(56.44\ \text{L})(2.00\ \text{atm})(281\ \text{K})}{(310\ \text{K})(23.52\ \text{L})} = 4.35\ \text{atm}$$

6.29 $$V_2 = \frac{P_1V_1T_2}{T_1P_2} = \frac{(756\ \text{mmHg})(30.0\ \text{mL})(260.5\ \text{K})}{(298\ \text{K})(325\ \text{mmHg})} = 61.0\ \text{mL}$$

6.31 (a) $$n = \frac{PV}{RT} = \frac{(1.33\ \text{atm})(50.3\ \text{L})}{(0.0821\ \text{L}\bullet\text{atm}\bullet\text{mol}^{-1}\bullet\text{K}^{-1})(350\ \text{K})} = 2.33\ \text{mol}$$

(b) The only information that we need to know about the gas is that it is an ideal gas.

6.33 Using the PV=nRT Gas Law equation, the following equation is derived:

$$MW = \frac{(\text{mass})RT}{PV} = \frac{(8.00\ \text{g})(0.0821\ \text{L}\bullet\text{atm}\bullet\text{mol}^{-1}\bullet\text{K}^{-1})(273\ \text{K})}{(2.00\ \text{atm})(22.4\ \text{L})} = 4.00\ \text{g/mol}$$

6.35 Using the PV=nRT equation, we can derive:

$$\frac{\text{mass}}{V} = \text{density} = \frac{P(MW)}{RT}$$

(a) At constant T, equation reduces to density = (constant) x (pressure), therefore, the density increases as pressure increases.
(b) At constant P, the equation reduces to density = (constant)(1/T), therefore, density decreases with increasing T.

6.37 Using the PV = nRT equation:

(a) $$n = \frac{PV}{RT} = \frac{(3.00\ \text{atm})(200\ \text{L})}{(0.0821\ \text{L}\bullet\text{atm}\bullet\text{mol}^{-1}\bullet\text{K}^{-1})(296\ \text{K})} = 24.7\ \text{mol O}_2$$

(b) $$\text{Mass of O}_2 = 24.7\ \text{mol O}_2\left(\frac{32.0\ \text{g O}_2}{1\ \text{mol O}_2}\right) = 790\ \text{g O}_2$$

6.39 $5.5\ \text{L air}\left(\frac{0.21\ \text{L } O_2}{1\ \text{L air}}\right) = 1.16\ \text{L } O_2$

$$\text{Moles of } O_2 = n = \frac{PV}{RT} = \frac{(1.1\ \text{atm})(5.5\ \text{L})}{(0.0821\ \text{L}\cdot\text{atm}\cdot\text{mol}^{-1}\cdot\text{K}^{-1})(310\ \text{K})} = 0.238\ \text{mol } O_2$$

$$0.238\ \text{mol } O_2\left(\frac{6.02\times10^{23}\ \text{molecules } O_2}{1\ \text{mol } O_2}\right) = 1.4\times10^{23}\ \text{molecules } O_2$$

6.41 1.0000 mole air = 0.7808 mol N_2 + 0.2095 mol O_2 + 0.0093 mol Ar
1.000 mole of air =

$$0.7808\ \text{mol}_{N_2}\left(\frac{28.01\ \text{g } N_2}{1\ \text{mol } N_2}\right) + 0.2095\ \text{mol}_{O_2}\left(\frac{32.00\ \text{g } O_2}{1\ \text{mol } O_2}\right) + 0.0093\ \text{mol}_{Ar}\left(\frac{39.95\ \text{g Ar}}{1\ \text{mol Ar}}\right)$$

(a) 1.000 mole of air = 28.95 g/mol

(b) $d(g/L)_{air} = \left(\frac{28.95\ \text{g air}}{1\ \text{mol air}}\right)\left(\frac{1\ \text{mol air}}{22.4\ \text{L}}\right) = 1.29\ \text{g/L}$

6.43 (a) $d_{SO_2} = \left(\frac{64.1\ \text{g}}{1\ \text{mol } SO_2}\right)\left(\frac{1\ \text{mol } SO_2}{22.4\ \text{L}}\right) = 2.86\ \text{g/L}$

(b) $d_{CH_4} = \left(\frac{16.0\ \text{g}}{1\ \text{mol } CH_4}\right)\left(\frac{1\ \text{mol } CH_4}{22.4\ \text{L}}\right) = 0.714\ \text{g/L}$

(c) $d_{H_2} = \left(\frac{2.02\ \text{g}}{1\ \text{mol } H_2}\right)\left(\frac{1\ \text{mol } H_2}{22.4\ \text{L}}\right) = 0.0902\ \text{g/L}$

(d) $d_{He} = \left(\frac{4.00\ \text{g}}{1\ \text{mol } H_2}\right)\left(\frac{1\ \text{mol He}}{22.4\ \text{L}}\right) = 0.179\ \text{g/L}$

(e) $d_{CO_2} = \left(\frac{44.0\ \text{g}}{1\ \text{mol } CO_2}\right)\left(\frac{1\ \text{mol } CO_2}{22.4\ \text{L}}\right) = 1.96\ \text{g/L}$

Gas comparison: SO_2 and CO_2 are denser than air; He, H_2 and CH_4 are less dense than air.

6.45 1.00 mL of octane (d = 0.7025 g/mL)

$$\text{Mass of octane} = 1.00 \text{ mL Octane}\left(\frac{0.7025 \text{ g Octane}}{1 \text{ mL Octane}}\right) = 0.7025 \text{ g Octane}$$

$$V_{Octane} = \frac{nRT}{P} = \frac{\left(\frac{0.7025 \text{ g}}{114.2 \text{ g} \cdot \text{mol}^{-1}}\right)(0.0821 \text{ L} \cdot \text{atm} \cdot \text{mol}^{-1} \cdot \text{K}^{-1})(373 \text{ K})}{(725 \text{ torr})\left(\frac{1.00 \text{ atm}}{760 \text{ torr}}\right)} = 0.197 \text{ L}$$

6.47 The densities would be the same. The density of a substance does not depend on its quantity.

6.49 (a) $P_T = P_{N2} + P_{O2} + P_{Ar}$

$P_{N_2} = (0.7808)(760 \text{ mm Hg}) = 594 \text{ mm Hg}$

$P_{O_2} = (0.2095)(760 \text{ mm Hg}) = 159 \text{ mm Hg}$

$P_{Ar} = (0.0093)(760 \text{ mm Hg}) = 7 \text{ mm Hg}$

$P_T = 760 \text{ mm Hg}$

(b) The total pressure exerted by the components is the sum of their partial pressures: 760 mm Hg (1.00 atm)

6.51 The total pressure should be the sum of its partial pressures.

$P_{N_2} = 560 \text{ mm Hg}$

$P_{O_2} = 210 \text{ mm Hg}$

$P_{CO_2} = 15 \text{ mm Hg}$

Partial Pressure Sum: 785 mm Hg $\quad P_T = 790 \text{ mm Hg}$

Difference of 5 mm Hg, therefore, there must be another gas present.

6.53 Covalent bonds are stronger than hydrogen bonds. Covalent bonds involve the sharing of electrons, where hydrogen bonds involve weaker electrostatic interactions.

6.55 Yes, the water OH can hydrogen bond (the hydrogen bond donor) with the oxygen lone pair on the S=O (the hydrogen bond acceptor).

$$(H_3C)_2S{=}\ddot{O}: \cdots H{-}O{-}H$$

6.57 Ethanol is a polar molecule and engages in intermolecular hydrogen bonding. Carbon dioxide is a nonpolar molecule and has only weak intermolecular London dispersion forces. The stronger hydrogen bonding intermolecular forces require more energy and higher temperatures to break before boiling.

6.59 Hexane has a higher boiling point. It is a larger molecule than butane, thus hexane has larger London dispersion forces to overcome before boiling.

6.61 Ionic compounds (salts) have the highest melting points because they require the greatest amount of energy to exceed the strongest of intermolecular forces (170-970 kcal/mol). An example includes sodium chloride (m.p. 801 °C). Nonpolar compounds have lowest melting points because London dispersion forces are the weakest of the intermolecular forces and require the least amount of energy to break (0.01-2 kcal/mol). An example of a nonpolar compound is naphthalene ($C_{10}H_8$, m.p. 80 °C).

6.63 $$39.2\text{ g }CF_2Cl_2\left(\frac{1\text{ mol }CF_2Cl_2}{120.9\text{ g }CF_2Cl_2}\right)\left(\frac{4.71\text{ kcal}}{1\text{ mol }CF_2Cl_2}\right) = 1.53\text{ kcal to vaporize the }CF_2Cl_2$$

6.65 (a) ~90 mm Hg (b) ~120 mm Hg (c) ~490 mm Hg

6.67 (a) HCl < HBr < HI
Increasing size of molecule increases London dispersion forces.

(b) O_2 < HCl < H_2O_2
O_2 has only weak London dispersion forces to overcome for boiling, where HCl is a polar molecule with stronger dipole-dipole attractions to overcome for boiling. H_2O_2 has the strongest intermolecular forces (hydrogen bonding) to exceed for boiling to occur.

6.69 The difference between heating water from 0°C to 37°C and heating ice from 0°C to 37°C is the heat of fusion.

The energy required to heat ice from 0°C to 37°C:

$$100 \text{ g } H_2O\left(\frac{1.0 \text{ cal}}{\text{g}\cdot{}^\circ\text{C}}\right)(37^\circ\text{C}) + 100 \text{ g } H_2O\left(\frac{80 \text{ cal}}{\text{g}}\right) = 11700 \text{ cal } (12 \text{ kcal})$$

The energy required to heat liquid water from 0°C to 37°C:

$$100 \text{ g } H_2O\left(\frac{1.0 \text{ cal}}{\text{g}\cdot{}^\circ\text{C}}\right)(37^\circ\text{C}) = 3700 \text{ cal } (3.7 \text{ kcal})$$

6.71 Sublimation is the conversion of a solid to gas, bypassing the liquid phase.

6.73 $$1.00 \text{ mL Freon-11}\left(\frac{1.49 \text{ g Freon-11}}{1 \text{ mL Freon-11}}\right)\left(\frac{1 \text{ mol Freon-11}}{137.4 \text{ g Freon-11}}\right) = 1.08 \times 10^{-2} \text{ mol Freon-11}$$

$$1.08 \times 10^{-2} \text{ mol Freon-11}\left(\frac{6.42 \text{ kcal}}{1 \text{ mol Freon-11}}\right) = 6.94 \times 10^{-2} \text{ kcal}$$

6.75 When the temperature of a substance increases, so does its entropy, therefore, a gas at 100°C has lower entropy than at 200°C.

6.77 When a person lowers their diaphragm, the volume of the chest cavity increases, thus lowering the pressure in the lungs relative to atmospheric pressure. Air at atmospheric pressure then rushes into the lungs, beginning the breathing process.

6.79 The first tapping sound one hears is the systolic pressure, which occurs when the sphygmomanometer pressure matches the blood pressure when the ventricle contracts, pushing blood into the arm.

6.81 When water freezes, it expands (one of the few substances that expands upon freezing) and will break the bottle when the ice expansion exceeds the volume of the bottle.

6.83 It is difficult to compress liquids and solids because their molecules or atoms are already very close together and there is very little empty space between them.

6.85 Conversion of psi to atm of an average tire pressure of 34 psi:

$$34 \text{ psi}\left(\frac{1 \text{ atm}}{14.7 \text{ psi}}\right) = 2.3 \text{ atm}$$

6.87 Aerosol cans already contain gases under high pressures. Gay-Lussac's Law predicts that the pressure inside the can will increase with increasing temperature, with the potential of the can explosively rupturing and causing injury.

6.89 $$V_2 = \frac{P_1V_1T_2}{T_1P_2} = \frac{(275\ \text{mm Hg})\left(\frac{1\ \text{atm}}{760\ \text{mm Hg}}\right)(387\ \text{mL})(378\ \text{K})}{(348\ \text{K})(1.36\ \text{atm})} = 112\ \text{mL}$$

6.91 Boiling point order:
H_2O (H-bonding) > $CHCl_3$ (dipole-dipole) > C_5H_{12} (London dispersion forces)

6.93 (a) As a gas is compressed under pressure, the molecules are forced closer together and the intermolecular forces pull the molecules together, forming a liquid.

(b) $$20\ \text{lbs propane}\left(\frac{1\ \text{kg}}{2.205\ \text{lb of propane}}\right) = 9.1\ \text{kg}$$

(c) $$9.1\ \text{kg propane}\left(\frac{1000\ \text{g propane}}{1\ \text{kg propane}}\right)\left(\frac{1\ \text{mole propane}}{44.1\ \text{g propane}}\right) = 2.1 \times 10^2\ \text{moles of propane}$$

(d) $$210\ \text{mol propane}\left(\frac{22.4\ \text{L propane}}{1\ \text{mol propane}}\right) = 4.7 \times 10^3\ \text{L propane}$$

6.95 $$d\ (\text{g/L}) = \left(\frac{0.00300\text{g}}{\text{cm}^3}\right)\left(\frac{1000\ \text{cm}^3}{\text{L}}\right) = 3.00\ \text{g/L}$$

$$MW = \frac{\text{massRT}}{\text{VP}} = \frac{(3.00\text{g})(0.0821\ \text{L}\cdot\text{atm}\cdot\text{mol}^{-1}\cdot\text{K}^{-1})(373\ \text{K})}{(1.00\ \text{L})(1.00\ \text{atm})} = 91.9\ \text{g/mol}$$

6.97 Use the PV = nRT equation after converting some of the units:

$$\text{Mol of } NH_3 = 60.0\ \cancel{g\,NH_3}\left(\frac{1\text{ mol }NH_3}{17.0\ \cancel{g\,NH_3}}\right) = 3.52\text{ mol }NH_3$$

$$\text{P (in atm)} = 77.2\ \cancel{\text{inch Hg}}\left(\frac{25.4\ \cancel{\text{mm Hg}}}{1\ \cancel{\text{inch Hg}}}\right)\left(\frac{1\text{ atm}}{760\ \cancel{\text{mm Hg}}}\right) = 2.58\text{ atm}$$

$$T = \frac{PV}{nR} = \frac{(2.58\ \cancel{\text{atm}})(35.1\ \cancel{L})}{(3.52\ \cancel{\text{mol}})(0.0821\ \cancel{L\cdot\text{atm}}\cdot\cancel{\text{mol}^{-1}}\cdot K^{-1})} = 313\text{ K }(40^{\circ}C)$$

6.99 The temperature of a liquid drops during evaporation because the molecules with higher kinetic energy leave the liquid as a gas, decreasing the average kinetic energy of the liquid. The temperature of the liquid is directly proportional to its average kinetic energy, therefore, the temperature decreases as the average kinetic energy decreases.

6.101 (a) Pressure on body = $100\ \cancel{ft}\left(\frac{1\text{ atm}}{33\ \cancel{ft}}\right) = 3.0\text{ atm}$

(b) At 1.00 atm, P_{N2} = 593 mm Hg (0.780 atm) and thus makes up 78.0% of the gas mixture, which does not change at the depth of 100 feet. At a depth of 100 feet, the total pressure on the lungs, which is equalized by pressure of air delivered by the SCUBA tank, is 3.0 atm.

$$P_{N_2}\text{ (at 100 ft)} = 3.0\ \cancel{\text{atm total pressure}}\left(\frac{0.78\text{ atm }P_{N_2}}{1\ \cancel{\text{atm total pressure}}}\right) = 2.34\text{ atm }P_{N_2}$$

(c) At 2 atm, P_{O2} = 158 mm Hg (0.208 atm) and thus makes up 20.8% of the gas mixture at 2 atm, which does not change at the depth of 100 feet. At a depth of 100 feet, the total pressure on the lungs, which is equalized by pressure of air delivered by the SCUBA tank, is 3.0 atm.

$$P_{O_2}\text{ (at 100 ft)} = 3.0\ \cancel{\text{atm total pressure}}\left(\frac{0.21\text{ atm }P_{O_2}}{1\ \cancel{\text{atm total pressure}}}\right) = 0.63\text{ atm }P_{O_2}$$

(d) As a diver ascends from 100 ft, the external pressure on the lungs decreases, therefore the volume of gasses in the lungs increases. If the diver does not exhale vigorously during a rapid ascent, the diver's lungs could over-inflate due to expanding gasses in the lungs, causing injury.

7.1 4.4% w/v KBr solution = 4.4 g KBr in 100 mL of solution

$$250\ \text{mL solution}\left(\frac{4.4\ \text{g KBr}}{100\ \text{mL solution}}\right) = 11\ \text{g KBr}$$

Add enough water to 11 g KBr to make 250 mL of solution

7.3 First, calculate the number of moles and mass of KCl that are needed:
Moles = M x V.

$$\text{Moles of KCl} = \left(\frac{1.06\ \text{mol KCl}}{1\ \text{L sol}}\right)(2.0\ \text{L sol}) = 2.12\ \text{mol KCl}$$

$$\text{Mass of KCl} = 2.12\ \text{mol KCl}\left(\frac{74.6\ \text{g KCl}}{1\ \text{mol KCl}}\right) = 158\ \text{g KCl}$$

Place 158 g of KCl into a 2-L volumetric flask, add some water, swirl until the solid has dissolved, and then fill the flask with water to the 2.0-L mark.

7.5 First, convert grams of glucose into moles of glucose, then convert moles of glucose into mL of solution:

$$\text{Moles of glucose} = 10.0\ \text{g glucose}\left(\frac{1\ \text{mol glucose}}{180\ \text{g glucose}}\right) = 0.0556\ \text{mol glucose}$$

$$0.0556\ \text{mol glucose}\left(\frac{1\ \text{L sol}}{0.300\ \text{mol glucose}}\right)\left(\frac{1000\ \text{mL sol}}{1\ \text{L sol}}\right) = 185\ \text{mL glucose sol}$$

7.7 Use the $M_1V_1 = M_2V_2$ equation:

$$V_1 = \frac{(0.600\ M\ \text{HCl})(300\ \text{mL sol HCl})}{(12.0\ M\ \text{HCl})} = 15.0\ \text{mL}$$

Place 15.0 mL of a 12.0 M HCl solution into a 300-mL volumetric flask, add some water, swirl until completely mixed, and then fill the flask with water to the 300-mL mark.

7.9 First calculate the mass of Na^+ ion in the 560 g of $NaHSO_4$. Then determine the Na^+ ion concentration in the solution using ppm.

$$560 \text{ g } NaHSO_4 \left(\frac{1 \text{ mol } NaHSO_4}{120 \text{ g } NaHSO_4}\right)\left(\frac{1 \text{ mol } Na^+}{1 \text{ mol } NaHSO_4}\right)\left(\frac{23.0 \text{ g } Na^+}{1 \text{ mol } Na^+}\right) = 107 \text{ g } Na^+$$

$$[Na^+] = \frac{107 \text{ g } Na^+}{4.5 \times 10^5 \text{ L sol}}\left(\frac{1 \text{ L sol}}{1 \text{ kg sol}}\right)\left(\frac{1 \text{ kg sol}}{1000 \text{ g sol}}\right) \times 10^6 \text{ ppm} = 0.24 \text{ ppm } Na^+$$

7.11 Compare the number of moles of ions or molecules in each solution. The solution with the most ions or molecules in solution will have the lowest freezing point.

Solution	Particle solution
(a) 6.2 *M* NaCl	2 x 6.2 *M* = 12.4 *M* ions
(b) 2.1 *M* $Al(NO_3)_3$	4 x 2.1 *M* = 8.4 *M* ions
(c) 4.3 *M* K_2SO_3	3 x 4.3 *M* = 12.9 *M* ions

Solution (c) has the highest concentration of solute particles (ions), therefore it will have the lowest freezing point.

7.13 The osmolarity of red blood cells is 0.30 osmol:

Solution	Particle solution
(a) 0.1 *M* Na_2SO_4	3 x 0.1 *M* = 0.3 osmol
(b) 1.0 *M* Na_2SO_4	3 x 1.0 *M* = 3.0 osmol
(c) 0.2 *M* Na_2SO_4	3 x 0.2 *M* = 0.6 osmol

Solution (a) has the same osmolarity as red blood cells, therefore is isotonic compared to red blood cells.

7.15 Glucose is being dissolved, therefore it is the solute. Water is dissolving the glucose, therefore water is the solvent.

7.17 (a) Wine (ethanol in water)
(b) Saline solution (NaCl dissolved in water)
(c) Carbonated water (carbon dioxide dissolved in water)
(d) Air (oxygen and nitrogen)

7.19 The prepared aspartic acid solution was unsaturated. Over two days time, some of the solvent (water) may have evaporated and the solution became supersaturated, precipitating the excess aspartic acid as a white solid.

7.21 (a) Ionic NaCl will dissolve in the polar water layer.
(b) Nonpolar camphor will dissolve in the nonpolar diethyl ether layer.
(c) Ionic KOH will dissolve in the polar layer.

7.23 Isopropyl alcohol would be a good first choice. The oil base in the paint is nonpolar. Both benzene and hexane are nonpolar solvents and may dissolve the paint, thus destroying the painting.

7.25 The solubility of aspartic acid at 25°C in 50.0 mL water is 0.250 g of solute. The cooled solution of 0.251 g aspartic acid in 50.0 mL water will be supersaturated by 0.001 g of aspartic acid.

7.27 According to Henry's Law, the solubility of a gas in a liquid is directly proportional to pressure. A closed bottle of a carbonated beverage is under pressure. After the bottle is opened, the pressure is released and the carbon dioxide becomes less soluble and escapes.

7.29 (a) Both quantities are equivalent to one significant figure.

$$\frac{1\ \cancel{\text{min}}}{2\ \cancel{\text{yr}}}\left(\frac{1\ \cancel{\text{yr}}}{365\ \cancel{\text{days}}}\right)\left(\frac{1\ \cancel{\text{day}}}{24\ \cancel{\text{hr}}}\right)\left(\frac{1\ \cancel{\text{hr}}}{60\ \cancel{\text{min}}}\right) \times 10^6\ \text{ppm} = 0.95\ \text{ppm}$$

$$\frac{1\ \cancel{\text{cent}}}{10{,}000\ \cancel{\text{dol}}}\left(\frac{1\ \cancel{\text{dol}}}{100\ \cancel{\text{cents}}}\right) \times 10^6\ \text{ppm} = 1\ \text{ppm}$$

(b) Both quantities are equivalent to one significant figure.

$$\frac{1\ \cancel{\text{min}}}{2000\ \cancel{\text{yr}}}\left(\frac{1\ \cancel{\text{yr}}}{365\ \cancel{\text{days}}}\right)\left(\frac{1\ \cancel{\text{day}}}{24\ \cancel{\text{hr}}}\right)\left(\frac{1\ \cancel{\text{hr}}}{60\ \cancel{\text{min}}}\right) \times 10^9\ \text{ppb} = 0.95\ \text{ppb}$$

$$\frac{1\ \cancel{\text{cent}}}{10{,}000{,}000\ \cancel{\text{dol}}}\left(\frac{1\ \cancel{\text{dol}}}{100\ \cancel{\text{cents}}}\right) \times 10^9\ \text{ppb} = 1\ \text{ppb}$$

7.31 (a) $\text{Vol. of ethanol} = 280\ \cancel{\text{mL}}\ \text{sol}\left(\frac{27\ \text{mL ethanol}}{100\ \cancel{\text{mL}}\ \text{sol}}\right) = 76\ \text{mL ethanol}$

76 mL ethanol dissolved in 204 mL water (to give 280 mL of solution)

(b) $\text{Vol. of ethyl acetate} = 435\ \cancel{\text{mL}}\ \text{sol}\left(\frac{1.8\ \text{mL ethyl acetate}}{100\ \cancel{\text{mL}}\ \text{sol}}\right) = 7.8\ \text{mL ethyl acetate}$

8 mL ethyl acetate dissolved in 427 mL water (to give 435 mL of solution)

(c) $\text{Vol. of benzene} = 1.65\ \cancel{\text{L}}\ \text{sol}\left(\frac{1000\ \cancel{\text{mL}}\ \text{sol}}{1\ \cancel{\text{L}}\ \text{sol}}\right)\left(\frac{8.00\ \text{mL benzene}}{100\ \cancel{\text{mL}}\ \text{sol}}\right) = 132\ \text{mL benzene}$

0.13 L benzene dissolved in 1.52 L chloroform (to give 1.65 L of solution)

7.33 (a) $\% \text{ (w/v)} = \frac{623 \text{ mg casein}}{15.0 \text{ mL sol}} \left(\frac{1 \text{ g casein}}{1000 \text{ mg casein}} \right) \times 100\% = 4.15\ \% \text{ w/v casein}$

(b) $\% \text{ (w/v)} = \frac{74 \text{ mg vit. C}}{250 \text{ mL sol}} \left(\frac{1 \text{ g vit. C}}{1000 \text{ mg vit. C}} \right) \times 100\% = 0.030\ \% \text{ w/v vitamin C}$

(c) $\% \text{ (w/v)} = \frac{3.25 \text{ g sucrose}}{186 \text{ mL sol}} \times 100\% = 1.75\ \% \text{ w/v sucrose}$

7.35 (a) $175 \text{ mL sol} \left(\frac{1 \text{ L sol}}{1000 \text{ mL sol}} \right) \left(\frac{1.14 \text{ mol } NH_4Br}{1 \text{ L solution}} \right) \left(\frac{97.9 \text{ g } NH_4Br}{1 \text{ mol } NH_4Br} \right) = 19.5 \text{ g } NH_4Br$

Place 19.5 g of NH_4Br into a 175-mL volumetric flask, add some water, swirl until completely dissolved, and then fill the flask with water to the 175-mL mark.

(b) $1.35 \text{ L sol} \left(\frac{0.825 \text{ mol NaI}}{1 \text{ L solution}} \right) \left(\frac{149.9 \text{ g NaI}}{1 \text{ mol NaI}} \right) = 167 \text{ g NaI}$

Place 167 g of NaI into a 1.35-L volumetric flask, add some water, swirl until completely dissolved, and then fill the flask with water to the 1.35-L mark.

(c) $330 \text{ mL sol} \left(\frac{1 \text{ L sol}}{1000 \text{ mL sol}} \right) \left(\frac{0.16 \text{ mol ethanol}}{1 \text{ L solution}} \right) \left(\frac{46.1 \text{ g ethanol}}{1 \text{ mol ethanol}} \right) = 2.4 \text{ g ethanol}$

Place 2.4 g of ethanol into a 330-mL volumetric flask, add some water, swirl until completely mixed, and then fill the flask with water to the 330-mL mark.

7.37 $M_{NaCl} = \frac{5.0 \text{ mg NaCl}}{0.5 \text{ mL sol}} \left(\frac{1 \text{ g NaCl}}{1000 \text{ mg NaCl}} \right) \left(\frac{1 \text{ mol NaCl}}{58.4 \text{ g NaCl}} \right) \left(\frac{1000 \text{ mL sol}}{1 \text{ L sol}} \right) = 0.2\ M \text{ NaCl}$

7.39 $M_{\text{glucose}} = \frac{22.0 \text{ g glucose}}{240 \text{ mL sol}} \left(\frac{1 \text{ mol glucose}}{180 \text{ g glucose}} \right) \left(\frac{1000 \text{ mL sol}}{1 \text{ L sol}} \right) = 0.509\ M \text{ glucose}$

$M_{K+} = \frac{190 \text{ mg } K^+}{240 \text{ mL sol}} \left(\frac{1 \text{ g}}{1000 \text{ mg}} \right) \left(\frac{1 \text{ mol } K^+}{39.1 \text{ g } K^+} \right) \left(\frac{1000 \text{ mL sol}}{1 \text{ L sol}} \right) = 0.0202\ M\ K^+$

$M_{Na+} = \frac{4.00 \text{ mg } K^+}{240 \text{ mL sol}} \left(\frac{1 \text{ g}}{1000 \text{ mg}} \right) \left(\frac{1 \text{ mol } Na^+}{23.0 \text{ g } Na^+} \right) \left(\frac{1000 \text{ mL sol}}{1 \text{ L sol}} \right) = 7.25 \times 10^{-4}\ M\ Na^+$

7.41 $$M_{sucrose} = \frac{13\ \text{g sucrose}}{15\ \text{mL sol}}\left(\frac{1\ \text{mol sucrose}}{342.3\ \text{g sucrose}}\right)\left(\frac{1000\ \text{mL sol}}{1\ \text{L sol}}\right) = 2.5\ M\ \text{sucrose}$$

7.43 Use the $\%_1V_1 = \%_2V_2$ equation:

$$V_2 = \frac{(0.750\%\ \text{w/v albumin})(5.00\ \text{mL sol})}{(0.125\%\ \text{w/v albumin})} = 30.0\ \text{mL}$$

The total volume of the dilution is 30.0 mL. Starting with a 5.00 mL solution, 25.0 mL of water must be added to reach a final volume of 30.0 mL.

7.45 Use the $\%_1V_1 = \%_2V_2$ equation:

$$V_1 = \frac{(0.25\%\ \text{w/v}\ H_2O_2)(250\ \text{mL sol})}{(30.0\%\ \text{w/v}\ H_2O_2)} = 2.1\ \text{mL}\ H_2O_2$$

Place 2.1 mL of 30.0% w/v H_2O_2 into a 250-mL volumetric flask, add some water, swirl until completely mixed, and then fill the flask with water to the 250-mL mark.

7.47 (a) $$\frac{12.5\ \text{mg Captopril}}{325\ \text{mg pill}} \times 10^6\ \text{ppm} = 3.85 \times 10^4\ \text{ppm Captopril}$$

(b) $$\frac{22\ \text{mg}\ Mg^{2+}}{325\ \text{mg pill}} \times 10^6\ \text{ppm} = 6.8 \times 10^4\ \text{ppm}\ Mg^{2+}$$

(c) $$\frac{0.27\ \text{mg}\ Ca^{2+}}{325\ \text{mg pill}} \times 10^6\ \text{ppm} = 8.3 \times 10^2\ \text{ppm}\ Ca^{2+}$$

7.49 Assume the density of the lake water to be 1.0 g/mL

$$1 \times 10^7\ \text{L water}\left(\frac{1000\ \text{mL water}}{1\ \text{L water}}\right)\left(\frac{1.0\ \text{g water}}{1.0\ \text{mL water}}\right) = 1 \times 10^{10}\ \text{g water}$$

$$\frac{0.1\ \text{g dioxin}}{1 \times 10^{10}\ \text{g water}} \times 10^9\ \text{ppb} = 0.01\ \text{ppb dioxin}$$

No, the dioxin level in the lake did not reach a dangerous level.

7.51 First, calculate the mass (in grams) of the nutrients in the cheese based on their percentages of daily allowances, then calculate concentration in ppm:

Iron: $(0.02)(15 \text{ mg Fe})\left(\dfrac{1 \text{ g Fe}}{1000 \text{ mg Fe}}\right) = 3 \times 10^{-4} \text{ g Fe}$

$$\frac{3 \times 10^{-4} \text{ g Fe}}{28 \text{ g cheese}} \times 10^{6} \text{ ppm} = 10 \text{ ppm Fe}$$

Calcium: $(0.06)(1200 \text{ mg Ca})\left(\dfrac{1 \text{ g Ca}}{1000 \text{ mg Ca}}\right) = 7 \times 10^{-2} \text{ g Ca}$

$$\frac{7 \times 10^{-2} \text{ g Ca}}{28 \text{ g cheese}} \times 10^{6} \text{ ppm} = 3 \times 10^{3} \text{ ppm Ca}$$

Vitamin A: $(0.06)(0.800 \text{ mg Vit. A})\left(\dfrac{1 \text{ g}}{1000 \text{ mg Vit. A}}\right) = 5 \times 10^{-5} \text{ g Vitamin A}$

$$\frac{5 \times 10^{-5} \text{ g Vit. A}}{28 \text{ g cheese}} \times 10^{6} \text{ ppm} = 2 \text{ ppm Vitamin A}$$

7.53 Both (a) 0.1 *M* KCl and (b) 0.1 *M* $(NH_4)_3PO_4$ will conduct electricity because they are strong electrolytes. A 0.1 *M* $(NH_4)_3PO_4$ solution (b) will have a greater conductivity than 0.1 *M* KCl because it has the largest concentration of dissociated ions (4 mole of ions dissociated per mole of $(NH_4)_3PO_4$, giving a 0.4 *M* solution of ions) compared to a 0.1 *M* KCl solution (2 mole of ions dissociated per mole of KCl, giving a 0.2 *M* solution of ions).

7.55 The polar compounds, (b), (c), and (d) will be soluble in water by forming hydrogen bonds with water.

7.57 (a) Homogeneous (b) Heterogeneous (c) Colloid
(d) Heterogeneous (e) Colloid (f) Colloid

7.59 As the temperature of the solution decreased, the protein molecules must have aggregated and formed a colloidal mixture. The turbid appearance is the result of the Tyndall effect.

7.61 $\Delta T = (1.86^{\circ}\text{C/mol})$(mole of particles in solution per 1000 g of water):

(a) $\Delta T = 1 \text{ mol NaCl}\left(\dfrac{2 \text{ mol particle}}{1 \text{ mol NaCl}}\right)\left(\dfrac{1.86^{\circ}\text{C}}{\text{mol particle}}\right) = 3.72^{\circ}\text{C}$ f.p. = -3.72°C

(b) $\Delta T = 1 \text{ mol MgCl}_2 \left(\frac{3 \text{ mol particle}}{1 \text{ mol MgCl}_2}\right)\left(\frac{1.86^\circ\text{C}}{\text{mol particle}}\right) = 5.58^\circ\text{C}$ f.p. = -5.58°C

(c) $\Delta T = 1 \text{ mol (NH}_4)_2\text{CO}_3 \left(\frac{3 \text{ mol particle}}{1 \text{ mol (NH}_4)_2\text{CO}_3}\right)\left(\frac{1.86^\circ\text{C}}{\text{mol particle}}\right) = 5.58^\circ\text{C}$

f.p. = -5.58°C

(d) $\Delta T = 1 \text{ mol Al(HCO}_3)_3 \left(\frac{4 \text{ mol particle}}{1 \text{ mol Al(HCO}_3)_3}\right)\left(\frac{1.86^\circ\text{C}}{\text{mol particle}}\right) = 7.44^\circ\text{C}$

f.p. = -7.44°C

7.63 Methanol is a non-electrolyte and does not dissociate. Use the following equation for freezing point depression while also converting to grams.

$$\text{Moles of CH}_3\text{OH per 1000 g H}_2\text{O} = \Delta T \left(\frac{1 \text{ mol particles}}{1.86^\circ\text{C}}\right)$$

$$20^\circ\text{C}\left(\frac{1 \text{ mol particles}}{1.86^\circ\text{C}}\right)\left(\frac{1 \text{ mole CH}_3\text{OH}}{1 \text{ mol particles}}\right)\left(\frac{32.0 \text{ g CH}_3\text{OH}}{1 \text{ mol CH}_3\text{OH}}\right) = 344 \text{ g CH}_3\text{OH}$$

7.65 Acetic acid is a weak acid, therefore does not completely dissociate into ions. KF is a strong electrolyte, completely dissociating into two ions and doubling the effect on freezing point depression compared to acetic acid.

7.67 In each case, the side with the greater osmolarity rises.
(a) B (b) B (c) A (d) B (e) A (f) same

7.69 (a) $\text{osmol}_{\text{Na}_2\text{CO}_3} = 0.39\ M \times 3 \text{ particles} = 1.2 \text{ osmol}$

(b) $\text{osmol}_{\text{Al(NO}_3)_3} = 0.62\ M \times 4 \text{ particles} = 2.5 \text{ osmol}$

(c) $\text{osmol}_{\text{LiBr}} = 4.2\ M \times 2 \text{ particles} = 8.4 \text{ osmol}$

(d) $\text{osmol}_{K_3PO_4} = 0.009\ M \times 4 \text{ particles} = 0.04 \text{ osmol}$

7.71 Cells in hypertonic solutions under crenation (shrink).

$$\text{osmol}_{\text{NaCl}} = \frac{0.9\ \text{g NaCl}}{100\ \text{mL sol}}\left(\frac{1000\ \text{mL sol}}{1\ \text{L sol}}\right)\left(\frac{1\ \text{mol NaCl}}{58.4\ \text{g NaCl}}\right)\left(\frac{2\ \text{mol particles}}{1\ \text{mol NaCl}}\right) = 0.3\ \text{osmol}$$

(a) $$\frac{0.3\ \text{g NaCl}}{100\ \text{mL sol}}\left(\frac{1000\ \text{mL sol}}{1\ \text{L sol}}\right)\left(\frac{1\ \text{mol NaCl}}{58.4\ \text{g NaCl}}\right)\left(\frac{2\ \text{mol particles}}{1\ \text{mol NaCl}}\right) = 0.1\ \text{osmol NaCl}$$

(b) $\text{osmol}_{\text{glucose}} = 0.9\ M \times 1\ \text{particle} = 0.9\ \text{osmol}$

(c) $$\frac{0.9\ \text{g glu}}{100\ \text{mL sol}}\left(\frac{1000\ \text{mL sol}}{1\ \text{L sol}}\right)\left(\frac{1\ \text{mol glu}}{180\ \text{g glu}}\right)\left(\frac{1\ \text{mol particles}}{1\ \text{mol glu}}\right) = 0.05\ \text{osmol glucose}$$

Solution (b) has a concentration greater than the isotonic solution so it will crenate red blood cells.

7.73 Carbon dioxide (CO_2) dissolves in normal rainwater to form a dilute solution of carbonic acid (H_2CO_3), which is a weak acid.

7.75 Nitrogen dissolved in the blood can lead to nitrogen narcosis, a narcotic effect also referred to as "rapture of the deep", which is similar to alcohol-induced intoxication.

7.77 The main component of limestone and marble is calcium carbonate ($CaCO_3$).

7.79 The minimum pressure required for the reverse osmosis in the desalinization of seawater exceeds 100 atm (the osmotic pressure of sea water).

7.81 $$\frac{0.2\ \text{g NaHCO}_3}{100\ \text{mL sol}}\left(\frac{1000\ \text{mL sol}}{1\ \text{L sol}}\right)\left(\frac{1\ \text{mol NaHCO}_3}{84.0\ \text{g NaHCO}_3}\right)\left(\frac{2\ \text{mol particles}}{1\ \text{mol NaHCO}_3}\right) = 0.05\ \text{osmol}$$

$\text{osmol}_{\text{NaHCO}_3} = 0.05\ \text{osmol}$

$$\frac{0.2\ \text{g KHCO}_3}{100\ \text{mL sol}}\left(\frac{1000\ \text{mL sol}}{1\ \text{L sol}}\right)\left(\frac{1\ \text{mol KHCO}_3}{100.1\ \text{g KHCO}_3}\right)\left(\frac{2\ \text{mol particles}}{1\ \text{mol KHCO}_3}\right) = 0.04\ \text{osmol}$$

$\text{osmol}_{\text{KHCO}_3} = 0.04\ \text{osmol}$

Yes, the change made a change in the tonicity. The error in replacing $NaHCO_3$ with $KHCO_3$ resulted in a hypotonic solution and an electrolyte imbalance by reducing the number of ions (osmolarity) in solution.

7.83 When the cucumber is placed in a saline solution, the osmolarity of the saline is greater than the water in the cucumber, so water moves from the cucumber to the saline solution. When a prune (partially dehydrated plum) is placed in the same solution, it expands because the osmolarity inside the prune is greater than the saline solution, so the water moves from saline solution to inside the prune.

7.85 The solubility of a gas is directly proportional to the pressure (Henry's Law) and inversely proportional to the temperature. The dissolved carbon dioxide formed a saturated solution in water when bottled at 2 atm of pressure. When the bottles are opened at atmospheric pressure, the gas becomes less soluble in the water. The carbon dioxide becomes supersaturated in water at room temperature and 1 atm , thus escapes through bubbles and frothing. In the other bottle, the solution of carbon dioxide in water is unsaturated at lower temperatures and does not lose carbon dioxide.

7.87 No, it would not be acceptable to use a 0.89% KCl solution for intravenous infusions because it will not be isotonic with blood. KCl has a higher molecular weight than NaCl, consequently its osmolarity will be smaller.

$$\frac{0.89\ \cancel{\text{g NaCl}}}{100\ \cancel{\text{mL}}\ \text{sol}}\left(\frac{1000\ \cancel{\text{mL sol}}}{1\ \text{L sol}}\right)\left(\frac{1\ \cancel{\text{mol NaCl}}}{58.4\ \cancel{\text{g NaCl}}}\right)\left(\frac{2\ \text{mol particles}}{1\ \cancel{\text{mol NaCl}}}\right) = 0.30\ \text{osmol NaCl}$$

$$\frac{0.89\ \cancel{\text{g KCl}}}{100\ \cancel{\text{mL}}\ \text{sol}}\left(\frac{1000\ \cancel{\text{mL sol}}}{1\ \text{L sol}}\right)\left(\frac{1\ \cancel{\text{mol KCl}}}{74.6\ \cancel{\text{g KCl}}}\right)\left(\frac{2\ \text{mol particles}}{1\ \cancel{\text{mol KCl}}}\right) = 0.24\ \text{osmol KCl}$$

7.89 $5.0\ \text{g reagent}\left(\frac{0.05\ \text{g Pb}}{10^{6}\ \text{g reagent}}\right) = 3 \times 10^{-7}\ \text{g Pb}$

7.91 $\text{osmol}_{\text{NaCl}} = \frac{0.9\ \cancel{\text{g NaCl}}}{100\ \cancel{\text{mL sol}}}\left(\frac{1000\ \cancel{\text{mL sol}}}{1\ \text{L sol}}\right)\left(\frac{1\ \cancel{\text{mol NaCl}}}{58.4\ \cancel{\text{g NaCl}}}\right)\left(\frac{2\ \text{mol particles}}{1\ \cancel{\text{mol NaCl}}}\right) = 0.3\ \text{osmol}$

$$\text{osmol}_{\text{dex}} = \frac{25\ \cancel{\text{g dex}}}{100\ \cancel{\text{mL sol}}}\left(\frac{1000\ \cancel{\text{mL sol}}}{1\ \text{L sol}}\right)\left(\frac{1\ \cancel{\text{mol dex}}}{15{,}000\ \cancel{\text{g dex}}}\right)\left(\frac{1\ \text{mol particles}}{1\ \cancel{\text{mol dex}}}\right) = 0.017\ \text{osmol}$$

The NaCl solution, solution (a), will have the greater osmotic pressure.

7.93 Use the $M_1V_1 = M_2V_2$ equation:

$$V_2 = \frac{M_1V_1}{M_2} = \frac{(1.18\ \cancel{\text{osmol}})(1.0\ \text{mL})}{0.30\ \cancel{\text{osmol}}} = 3.9\ \text{mL (final volume)}$$

3.9 mL - 1.0 mL = 2.9 mL of H_2O added to sea water (1.0 mL) to reach blood osmolarity

7.95 The ethanol displaced the water from the solvation layer of the hyaluronic acid and thus allowed hyaluronic acid molecules to stick together upon collision and aggregate.

8.1 The rate of the reaction is equal to the change in the amount of O_2 per unit time.

$$\text{Rate of } O_2 \text{ formation} = \frac{(0.35 \text{ L } O_2 - 0.020 \text{ L } O_2)}{15 \text{ min}} = 0.022 \text{ L } O_2/\text{min}$$

8.3 $$K = \frac{[H_2SO_4]}{[SO_3][H_2O]}$$

8.5 $$K = \frac{[PCl_5]}{[PCl_3][Cl_2]} = \frac{[1.66\ M]}{[1.66\ M][1.66\ M]} = 0.602\ M^{-1}$$

8.7 Le Chatelier's principle would predict that the equilibrium would shift to the left by adding a product (Br_2).

8.9 If the equilibrium shifts right with the addition of heat, heat must have been a reactant and the reaction endothermic.

8.11 $$\text{Rate of } CH_3I \text{ formation} = \frac{(0.840\ M\ CH_3I - 0.260\ M\ CH_3I)}{80 \text{ min}} = 7.25 \times 10^{-3}\ M\ CH_3I/\text{min}$$

8.13 Reactions involving aqueous solution of ions require no bond breaking and are very fast because of low activation energies. Reactions between covalent molecules require covalent bonds to be broken, requiring higher activation energies, thus slower reaction rates.

8.15 The following energy diagram can be drawn for an exothermic reaction:

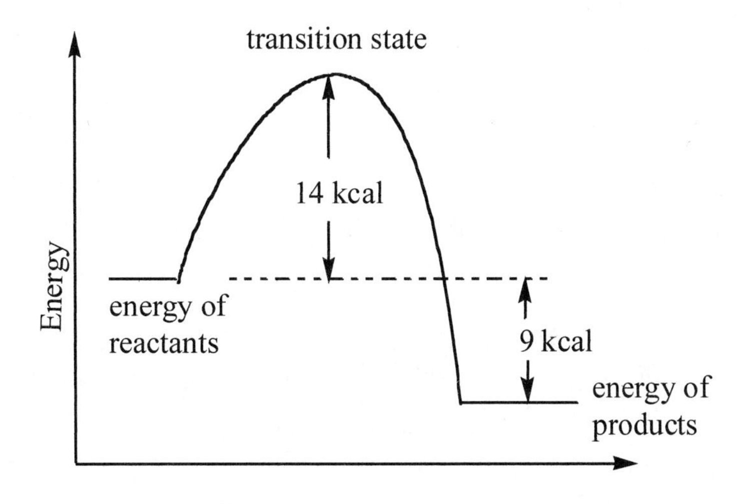

ule for temperature effect on reaction rates states that for every temperature
0 °C, the reaction rate doubles.

	10°C	→	20°C	→	30°C	→	40°C	→	50°C
	16 hr	→	8 hr	→	4 hr	→	2 hr	→	1 hr

A reaction temperature of 50 °C corresponds to a reaction completion time of 1 hr.

8.19 (1) Increase the temperature
(2) Increase the concentration of reactants
(3) Add a catalyst

8.21 A catalyst increases the rate by providing an alternate reaction pathway of lower activation energy.

8.23 Examples of irreversible reactions include: digesting a piece of candy, the rusting of iron, exploding TNT, and the reaction of sodium or potassium with water.

8.25 (a) $K = \frac{[H_2O]^2[O_2]}{[H_2O_2]^2}$

(b) $K = \frac{[N_2O_4]^2[O_2]}{[N_2O_5]^2}$

(c) $K = \frac{[C_6H_{12}O_6][O_2]^6}{[H_2O]^6[CO_2]^6}$

8.27 $K = \frac{[CO_2][H_2]}{[H_2O][CO]} = \frac{[0.133\ M][3.37\ M]}{[0.720\ M][0.933\ M]} = 0.667$

8.29 $K = \frac{[NO]^2[Cl_2]}{[NOCl_2]^2} = \frac{[1.4\ M]^2[0.34\ M]}{[2.6\ M]^2} = 0.099\ M$

8.31 When K > 1, equilibrium favors products; when K < 1, equilibrium favors reactants. Products are favored in (b) and (c). Reactants are favored in (a), (d), and (e)

8.33 No, the rate of reaction is independent of the energy difference between products and reactant. The rate of reaction is inversely proportional to the activation energy.

8.35 The reaction reaches equilibrium quickly, but the equilibrium favors the reactants ($K < 1$) and would not be a very good industrial process.

8.37 (a) Right (b) Right (c) Left (d) Left (e) No shift

8.39 (a) Adding Br_2 (a reactant), will shift the equilibrium to the right.
(b) The equilibrium constant will remain the same.

8.41 (a) No change (b) No change (c) Smaller
Equilibrium constants are independent of reactant and product concentrations. The K of an endothermic reaction will decrease with decreasing temperature

8.43 As temperatures increase, the rates of most chemical processes increase. A high body temperature is dangerous because metabolic processes (including digestion, respiration, and the biosynthesis of essential compounds) take place at a rate faster than what is safe for the body. As temperatures decrease, so do the rates of most chemical reactions. As body temperatures decrease below normal, the vital chemical reactions will slow down to rates slower than what is safe for the body.

8.45 The capsule with the tiny beads will act faster than the solid pill form. The small bead size increases the drug's surface area allowing the drug to react faster and deliver its therapeutic effects more quickly.

8.47 The addition of heat is used to increase the rate of reaction. The addition of a catalyst permits the reaction to take place at a convenient rate and temperature.

8.49 The following energy diagram can be drawn for an exothermic reaction:

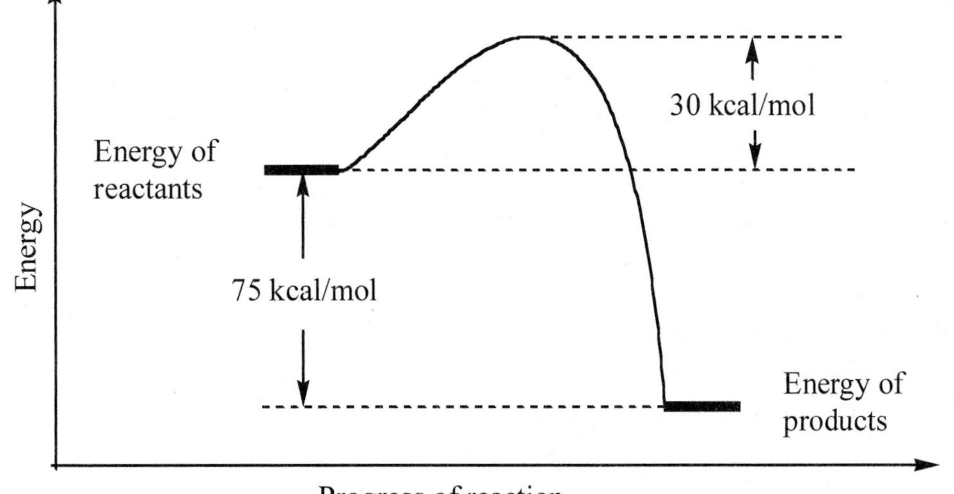

8.51 Rate = k[NOBr]

$$k = \frac{\text{rate}}{[\text{NOBr}]} = \frac{-2.3 \text{ mol NOBr}/\cancel{L} \cdot \text{hr}}{6.2 \text{ mol NOBr}/\cancel{L}} = -0.37/\text{hr}$$

8.53 $$\text{Rate} = \frac{[0.180 \text{ mol/L } N_2O_4 - 0.200 \text{ mol/L } N_2O_4]}{10 \text{ s}} = -2.0 \times 10^{-3} \text{ mol } N_2O_4 / \text{L} \cdot \text{s}$$

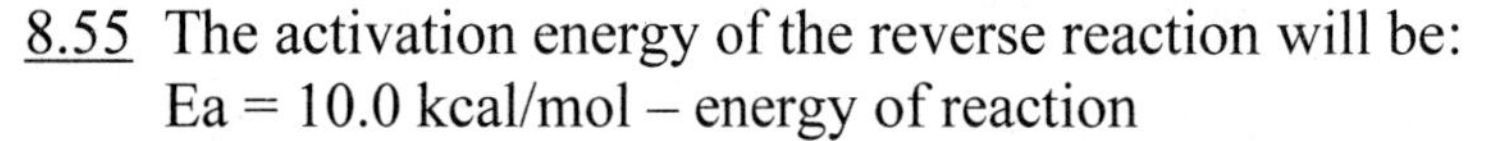

8.55 The activation energy of the reverse reaction will be:
Ea = 10.0 kcal/mol – energy of reaction

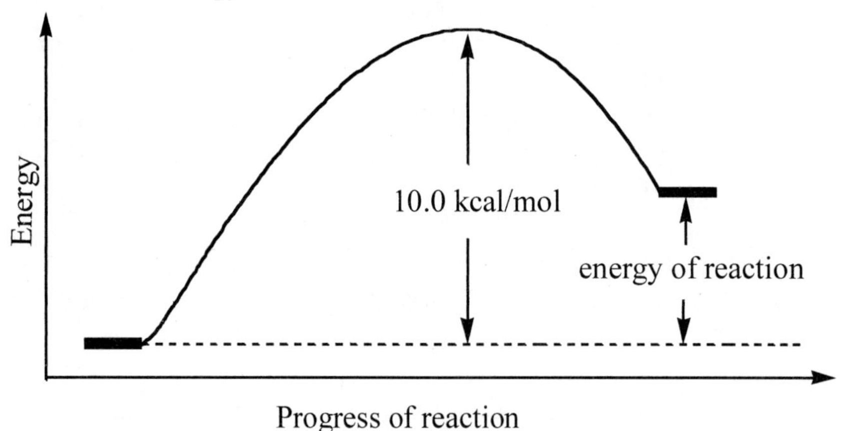

8.57 The temperature increase is 20°C, and the rate doubles twice. The rate of reaction at 320 K is 0.88 *M*/min.

8.59 Endothermic reactions must have an activation energy greater than zero. By definition, endothermic reactions occur where the products have a greater energy than the reactants.

8.61 Assuming that there is excess AgCl in the previous recipe, the new recipe does not need to change. The desert conditions add nothing that would affect an equilibrium initiated by light.

8.63 (a) At equilibrium: [Ethanol] = 0.33 mol and [Ethyl acetate] = [H_2O] = 0.67 mol.

(b) $$K_{eq} = \frac{[\text{Ethyl acetate}][H_2O]}{[\text{Acetic acid}][\text{Ethanol}]} = \frac{[0.33 \text{ mol}[0.33\text{mol}]}{[0.67 \text{ mol}][0.67\text{mol}]} = 0.24$$

9.1 Acid reaction for HPO_4^{2-}:

$$HPO_4^{2-} + H_2O \rightleftharpoons H_3O^+ + PO_4^{3-}$$

Base reaction for HPO_4^{2-}:

$$HPO_4^{2-} + H_3O^+ \rightleftharpoons H_2O + H_2PO_4^{1-}$$

9.3 $pK_a = -\log K_a$
pK_a HCN $= -\log 4.9 \times 10^{-10} = 9.31$

9.5 $[H_3O^+][OH^-] = 1 \times 10^{-14}$

$$[H_3O^+] = \frac{1.0 \times 10^{-14}}{1.0 \times 10^{-12}} = 1.0 \times 10^{-2}\ M$$

9.7 pH + pOH = 14
$[OH^-] = 1.0 \times 10^{-4}$ pOH = 4
$[H_3O^+] = 1.0 \times 10^{-10}$ pH = 10

9.9 The pH of a buffer solution containing equimolar quantities of acid and conjugate base is equal to the pK_a of the weak acid.
(a) NH_4Cl and NH_3: pK_a $NH_4Cl = 9.25$
(b) CH_3COOH and CH_3COONa: pK_a $CH_3COOH = 4.75$

9.11 Use the Henderson-Hasselbalch Equation: $pH = pK_a + \log[A^-]/[HA]$ where:

$$pH = 8.3 + \log\left(\frac{0.05\text{ mol}}{0.2\text{ mol}}\right) = 7.7$$

The mole amount of TRIS acid and base was used instead of concentration because they are both dissolved in 500 mL of water, therefore, the amount of water cancels out. The TRIS buffer will have a pH of 8 (to one significant figure).

9.13 Listed below are the following acid ionization equilibrium equations:

(a) $HNO_3(aq) + H_2O(l) \rightleftharpoons H_3O^+(aq) + NO_3^-(aq)$

(b) $HBr(g) + H_2O(l) \rightleftharpoons H_3O^+(aq) + Br^-(aq)$

(c) $H_2SO_3(aq) + H_2O(l) \rightleftharpoons H_3O^+(aq) + HSO_3^-(aq)$

(d) $H_2SO_4(aq) + H_2O(l) \rightleftharpoons H_3O^+(aq) + HSO_4^-(aq)$

(e) $HCO_3^-(aq) + H_2O\ (l) \rightleftharpoons H_3O^+(aq) + CO_3^{2-}(aq)$

(f) $H_3BO_3(aq) + H_2O(l) \rightleftharpoons H_3O^+(aq) + H_2BO_3^-(aq)$

9.15 Acids (a), (c), (e), and (f) have pK_a's > 0, therefore do not ionize completely and are considered to be weak acids. Acids (b) and (d) have pK_a's < 0; therefore are ionized completely and are considered to be strong acids.

9.17 Acids that have pK_a's > 0 do not ionize completely and are considered to be weak acids; therefore, an acid with a pK_a of 2.1 is a weak acid.

9.19 (a) Brønsted-Lowry acids are considered proton donors.
(b) Brønsted-Lowry bases are considered proton acceptors.

9.21 A conjugate base is the species that results after the acid lost its proton.
(a) HPO_4^{2-} (b) HS^- (c) CO_3^{2-}
(d) $CH_3CH_2O^-$ (e) OH^-

9.23 A conjugate acid results from a base acquiring a proton from an acid:
(a) H_3O^+ (b) $H_2PO_4^-$ (c) $CH_3NH_3^+$
(d) HPO_4^{2-} (e) $H_2BO_3^-$

9.25 The equilibrium favors the side with the weaker acid-weaker base.
Equilibria (b) and (c) favor the left, equilibrium (a) favors the right.

(a) $C_6H_5OH + C_2H_5O^- \rightleftharpoons C_6H_5O^- + C_2H_5OH$
stronger acid, stronger base, weaker base, weaker acid

(b) $HCO_3^- + H_2O \rightleftharpoons H_2CO_3 + OH^-$
weaker base, weaker acid, stronger acid, stronger base

(c) $CH_3COOH + H_2PO_4^- \rightleftharpoons CH_3COO^- + H_3PO_4$
weaker acid, weaker base, stronger base, stronger acid

9.27 (a) Strong acids have smaller pK_a's, therefore weak acids have large pK_a's.
(b) Strong acids have large K_a's.

9.29 At equal concentrations, pH decreases (becomes more acidic) as K_a increases.
(a) 0.10 *M* HCl (b) 0.10 *M* H_3PO_4 (c) 0.010 *M* H_2CO_3
(d) 0.10 *M* NaH_2PO_4 (e) 0.10 *M* Aspirin

9.31 Only (b) Mg involves a redox reaction. The other reactions are acid-base reactions.

(a) $Na_2CO_3 + 2HCl \rightarrow CO_2 + 2NaCl + H_2O$

(b) $Mg + 2HCl \rightarrow MgCl_2 + H_2$

(c) $NaOH + HCl \rightarrow NaCl + H_2O$

(d) $Fe_2O_3 \quad 6HCl \rightarrow 2FeCl_3 + 3H_2O$

(e) $NH_3 + HCl \rightarrow NH_4Cl$

(f) $CH_3NH_2 + HCl \rightarrow CH_3NH_3Cl$

(g) $NaHCO_3 + HCl \rightarrow H_2CO_3 + NaCl \rightarrow CO_2 + H_2O + NaCl$

9.33 Using the equation: $[H_3O^+][OH^-] = 1.0 \times 10^{-14}\ M^2$

(a) $[OH^-] = 10^{-3}\ M$ (b) $[OH^-] = 10^{-10}\ M$

(c) $[OH^-] = 10^{-7}\ M$ (d) $[OH^-] = 10^{-15}\ M$

9.35 Using the equation: $pH = -\log[H_3O^+]$:

(a) pH = 8 (basic) (b) pH = 10 (basic) (c) pH = 2 (acidic)

(d) pH = 0 (acidic) (e) pH = 7 (neutral)

9.37 Using the equation: $pH = -\log[H_3O^+]$ and pH + pOH = 14.

(a) pH = 8.5 (basic) (b) pH = 1.2 (acidic)

(c) pH = 11 (basic) (d) pH = 6.3 (acidic)

9.39 Using the equation: $[OH^-] = 10^{-pOH}$ and pH + pOH = 14.

(a) pOH = 1.0, $[OH^-] = 0.10\ M$ (b) pOH = 2.4, $[OH^-] = 4.0 \times 10^{-3}\ M$

(c) pOH = 2.0, $[OH^-] = 1.0 \times 10^{-2}\ M$ (d) pOH = 5.6, $[OH^-] = 2.5 \times 10^{-6}\ M$

9.41 $$M = \frac{\text{mol}}{\text{L}} = \left(\frac{12.7\ \text{g HCl}}{1.00\ \text{L sol}}\right)\left(\frac{1\ \text{mol HCl}}{36.5\ \text{g HCl}}\right) = 0.348\ M\ \text{HCl}$$

9.43 (a) 12 g of NaOH diluted to 400 mL of solution:

$$400\ \text{mL sol}\left(\frac{1\ \text{L sol}}{1000\ \text{mL sol}}\right)\left(\frac{0.75\ \text{mol NaOH}}{1\ \text{L sol}}\right)\left(\frac{40.0\ \text{g NaOH}}{1\ \text{mol NaOH}}\right) = 12\ \text{g NaOH}$$

(b) 12 g of $Ba(OH)_2$ diluted to 1.0 L of solution:

$$\frac{0.071\ \text{mol Ba(OH)}_2}{1\ \text{L sol}}\left(\frac{171.4\ \text{g Ba(OH)}_2}{1\ \text{mol Ba(OH)}_2}\right) = 12\ \text{g Ba(OH)}_2$$

9.45 5.66 mL of 0.740 M H_2SO_4 are required to titrate 27.0 mL of 0.310 M NaOH.

$$27.0 \text{ mL NaOH sol}\left(\frac{1 \text{ L sol}}{1000 \text{ mL sol}}\right)\left(\frac{0.310 \text{ mol NaOH}}{1 \text{ L sol}}\right) = 8.37\times10^{-3} \text{ mol NaOH}$$

$$8.37\times10^{-3} \text{ mol NaOH}\left(\frac{1 \text{ mol } H_2SO_4}{2 \text{ mol NaOH}}\right)\left(\frac{1 \text{ L } H_2SO_4 \text{ sol}}{0.740 \text{ mol } H_2SO_4}\right)\left(\frac{1000 \text{ mL sol}}{1 \text{ L sol}}\right) = 5.66 \text{ mL}$$

9.47 Assuming that the base generates one mole of hydroxide per mole of base:

$$22.0 \text{ mL HCl sol}\left(\frac{0.150 \text{ mol HCl}}{1000 \text{ mL sol}}\right)\left(\frac{1 \text{ mol } H^+}{1 \text{ mol HCl}}\right) = 3.30 \times 10^{-3} \text{ mol } H^+$$

At the end point, 3.30×10^{-3} mol of H^+ added to 3.30×10^{-3} mol of the unknown base.

9.49 The point at which an indicator changes color is called the end point.

9.51 In the CH_3COOH/CH_3COO^- buffer solution, the CH_3COO^- is completely ionized while the CH_3COOH is only partially ionized.

(a) $H_3O^+ + CH_3COO^- \rightleftharpoons CH_3COOH + H_2O$ (removal of H_3O^+)

(b) $HO^- + CH_3COOH \rightleftharpoons CH_3COO^- + H_2O$ (removal of OH^-)

9.53 Yes, the conjugate acid becomes the weak acid and the weak base becomes the conjugate base.

9.55 The pH of a buffer can be changed by altering the weak acid/conjugate base ratio according to the Henderson-Hasselbalch equation. The buffer capacity can be changed without change in pH by increasing or decreasing the amount of weak acid/conjugate base mixture while keeping the ratio of the two constant.

9.57 This would occur in a couple of cases. One is very common, which is where you are using a buffer, such as TRIS with a pKa of 8.3, but you do not want the solution to have a pH of 8.3. If you wanted a pH of 8.0, for example, you would have to have unequal amounts of the conjugate acid and base, with there being more conjugate acid. Another might be a situation where you are performing a reaction that you know will generate H^+ but you want the pH to be stable. In that situation, you might start with a buffer that was initially set to have more of the conjugate base so that it could absorb more of the H^+ that you know will be produced.

9.59 (a) According to the Henderson-Hasselbalch equation, no change in pH will be observed as long as the ratio of weak acid/conjugate base ratio remains the same.
(b) The buffer capacity increases with increasing amount of weak acid/conjugate base concentrations, therefore, 1.0 mol amounts of each diluted to 1 L would have a greater buffer capacity than 0.1 mol of each diluted to 1 L.

9.61 Using the Henderson-Hasselbalch Equation:

(a) $$pH = pK_a + \log\frac{[\text{lactate}^-]}{[\text{lactic acid}]} = 3.85 + \log\frac{[0.40\ M]}{[0.80\ M]} = 3.5$$

(b) $$pH = pK_a + \log\frac{[NH_3]}{[NH_4^+]} = 9.25 + \log\frac{[0.30\ M]}{[1.50\ M]} = 8.6$$

9.63 The 100 mL of 0.1 *M* phosphate buffer initially contains 0.01 moles of NaH_2PO_4 and Na_2HPO_4 components.

If the two components of the buffer are equal: $pH = 7.21 + \log\frac{[0.1M]}{[0.1M]} = 7.21$

A pH of 6.8 suggests that there is more of the acid (greater than 0.01 moles of NaH_2PO_4) than conjugate base (less than 0.01 moles of Na_2HPO_4). The addition of 10 mL of 1 *M* HCl (0.01 mol) will overwhelm the buffers' capacity to neutralize the additional acid.

9.65 $$pH = pK_a + \log\frac{[\text{TRIS}]}{[\text{TRIS-H}]^+}$$

$$pH = 8.3 + \log\frac{[0.05M]}{[0.1M]} = 8.0$$

9.67 No. HEPES has a pK_a of 7.55, which means it is a useable buffer between pH 6.55 and 8.55.

9.69 $Mg(OH)_2$ is a weak base used as a flame-retardant in plastics.

9.71 Both strong acids and strong bases are very harmful to the eyes, but strong bases are more harmful to the cornea because the healing of the wounds can deposit non-transparent scar tissue that impairs vision.

9.73 (a) Respiratory acidosis is caused by hypoventilation, which is caused by a variety of breathing difficulties, such as a windpipe obstruction, asthma, or pneumonia.
(b) Metabolic acidosis is caused by starvation or heavy exercise.

9.75 Sodium bicarbonate is the weak base form of one of the blood buffers. It will tend to raise the pH of the blood, which is the purpose of the sprinter's trick, so that the person can absorb more H^+ during the event. By putting $NaHCO_3$ into the system, the following reaction will occur:

$$HCO_3^-(aq) + H^+(aq) \rightleftharpoons H_2CO_3(aq)$$

The loss of the H^+ means that the blood pH will rise.

9.77 The equilibrium favors the side of the weaker acid/weaker base.

(a) Benzoic acid is soluble in aqueous NaOH.

$$C_6H_5COOH + NaOH \rightleftharpoons C_6H_5COO^- + H_2O$$

$pK_a = 4.19$ $\qquad$ $pK_a = 15.56$

(b) Benzoic acid is soluble in aqueous $NaHCO_3$.

$$C_6H_5COOH + NaHCO_3 \rightleftharpoons CH_3C_6H_4O^- + H_2CO_3$$

$pK_a = 4.19$ $\qquad$ $pK_a = 6.37$

(c) Benzoic acid is soluble in aqueous Na_2CO_3.

$$C_6H_5COOH + CO_3^{2-} \rightleftharpoons CH_3C_6H_4O^- + HCO_3^-$$

$pK_a = 4.19$ $\qquad$ $pK_a = 10.25$

9.79 The strength of an acid is not important to the amount of NaOH that would be required to hit a phenolphthalein endpoint. Therefore, the more concentrated acid, the acetic acid, would require more NaOH.

9.81 The solution of oxalic acid is 3.70×10^{-3} *M*

$$\frac{0.583 \text{ g } \cancel{H_2C_2O_4}}{1.75 \text{ L sol}}\left(\frac{1 \text{ mol } H_2C_2O_4}{90.04 \text{ g } \cancel{H_2C_2O_4}}\right) = 3.70 \times 10^{-3} \text{ M oxalic acid}$$

9.83 The concentration of barbituric acid equilibrates to 0.90 *M*.

$$K_a = \frac{[\text{Barbiturate}^-][H_3O^+]}{[\text{Barbituric acid}]} \qquad x = [H_3O^+] = [\text{Barbiturate}^-]$$

$$[\text{Barbituric acid}] = \frac{x^2}{K_a} = \frac{(0.0030)^2}{1.0 \times 10^{-5}} = 0.90\ M$$

9.85 Yes, a pH = 0 is possible. A 1.0 *M* solution of HCl has a $[H_3O^+] = 1.0$ *M*.

$$pH = -\log[H_3O^+] = -\log[1.0\ M] = 0$$

9.87 The qualitative relationship between acids and their conjugate bases states that the stronger the acid, the weaker its conjugate base. This can be quantified in the equation: $K_b \times K_a = K_w$ or $K_b = 1.0 \times 10^{-14}/K_a$ where K_b is the base dissociation equilibrium constant for the conjugate base, K_a is the acid dissociation equilibrium constant for the acid, and K_w is the ionization equilibrium constant for water.

9.89 Yes. The strength of the acid is irrelevant. Both acetic acid and HCl have one H^+ to give up, so equal moles of either will require equal moles of NaOH to titrate to an endpoint.

9.91 Using the Henderson-Hasselbalch equation:

$$\frac{[H_2BO_3^-]}{[H_3BO_3]} = 10^{pH - pKa} = 10^{8.40-9.14}$$

$$\frac{[H_2BO_3^-]}{[H_3BO_3]} = 0.182$$

Need 0.182 mol of $H_2BO_3^-$ and 1.00 mol of H_3BO_3 in 1.00 L of solution.

9.93 Equilibria favor the side of the weaker acid/weaker base. Large pK_a values correlate with weak acids and small pK_a values correlate with strong acids, therefore, equilibria favor the side with the largest pK_a values.

9.95 (a) $HCOO^- + H_3O^+ \rightleftharpoons HCOOH + H_2O$

(b) $HCOOH + HO^- \rightleftharpoons HCOO^- + H_2O$

9.97 Using the Henderson-Hasselbalch Equation:

(a) $[Na_2HPO_4] = [NaH_2PO_4]10^{pH - pKa} = [0.050M]10^{7.21-7.21} = 0.050\ M$

(b) $[Na_2HPO_4] = [NaH_2PO_4]10^{pH - pKa} = [0.050M]10^{6.21-7.21} = 0.0050\ M$

(c) $[Na_2HPO_4] = [NaH_2PO_4]10^{pH - pKa} = [0.050M]10^{8.21-7.21} = 0.50\ M$

9.99 According to the Henderson-Hasselbalch equation:

$$pH = 7.21 + \log\frac{[HPO_4^{2-}]}{[H_2PO_4^-]}$$

As the concentration of $H_2PO_4^-$ increases, the $\log\frac{[HPO_4^{2-}]}{[H_2PO_4^-]}$ becomes negative,

thus lowering the pH and becoming more acidic.

9.101 No. A buffer will only have a pH equal to its pK_a if there are equimolar amounts of the conjugate acid and base forms. If this is the basic form of TRIS, then just putting any amount of that into water will give a pH much higher than the pK_a value.

9.103 (a) pH = 7.1, $[H_3O^+] = 7.9 \times 10^{-8}$ *M*, basic
(b) pH = 2.0, $[H_3O^+] = 7.9 \times 10^{-2}$ *M*, acidic
(c) pH = 7.4, $[H_3O^+] = 4.0 \times 10^{-8}$ *M*, basic
(d) pH = 7.0, $[H_3O^+] = 1.0 \times 10^{-7}$ *M*, neutral
(e) pH = 6.6, $[H_3O^+] = 2.5 \times 10^{-7}$ *M*, acidic
(f) pH = 7.4, $[H_3O^+] = 4.0 \times 10^{-8}$ *M*, basic
(g) pH = 6.5, $[H_3O^+] = 3.2 \times 10^{-7}$ *M*, acidic
(h) pH = 6.9, $[H_3O^+] = 1.3 \times 10^{-7}$ *M*, acidic

9.105 Using the Henderson-Hasselbalch equation:

$$7.9 = 7.21 + \log\frac{[HPO_4^{2-}]}{[H_2PO_4^-]}$$

$$\frac{[HPO_4^{2-}]}{[H_2PO_4^-]} = 10^{0.19} = 4.9$$

10.1 Following are Lewis structures showing all bond angles.

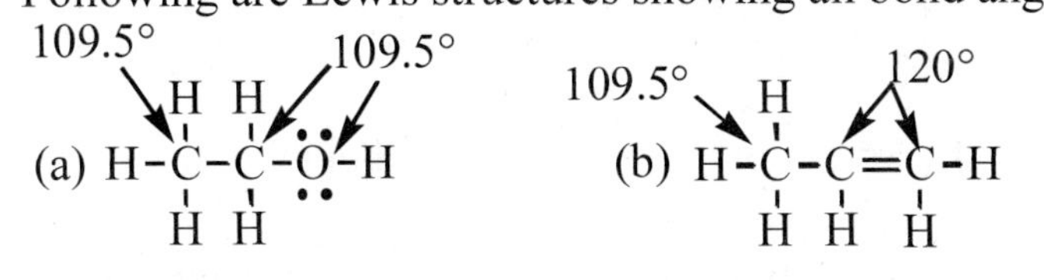

10.3 The three secondary (2°) amines with the molecular formula $C_4H_{11}N$ are:

$CH_3CH_2CH_2NHCH_3$ $\quad$ $CH_3CH(CH_3)NHCH_3$ $\quad$ $CH_3CH_2NHCH_2CH_3$

10.5 The two carboxylic acids with the molecular formula $C_4H_8O_2$ are:

$CH_3CH_2CH_2C(=O)OH$ $\quad$ $CH_3CH(CH_3)C(=O)OH$

10.7 Assuming that each vanillin is pure, there is no difference.

10.9 Wöhler heated ammonium chloride and silver cyanate, both inorganic compounds, and obtained urea, an organic compound.

10.11 Among the textile fibers, think of natural fibers such as cotton, wool, and silk. Also, think of synthetic textile fibers such as Nylon, Dacron polyester, and polypropylene.

10.13 For hydrogen, the number of valence electrons plus the number of bonds equals 2. For carbon, nitrogen, and oxygen, the number of valence electrons plus the number of bonds equals 8. Carbon, with four valence electrons, forms four bonds. Nitrogen, with five valence electrons, forms three bonds. Oxygen, with six valence electrons, forms two bonds.

10.15 Following are Lewis structures for each compound.

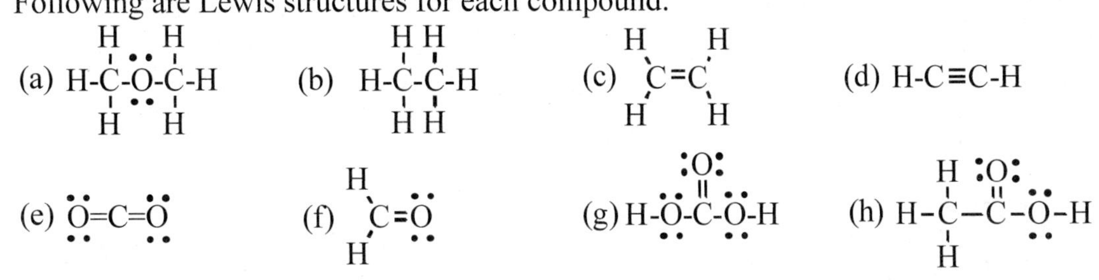

10.17 In stable organic compounds, carbon must have four covalent bonds to other atoms. In (a), carbon has bonds to five other atoms. In (b), one carbon has four bonds to other atoms, but the second carbon has five bonds to other atoms.

10.19 You would find three regions of electron density and, therefore, predict 120° for the H-N-H bond angles.

10.21 (a) 109.5° about C and O (b) 120° about C (c) 180° about C

10.23 A functional group is a part of an organic molecule that may undergo a chemical reaction(s) depending on the conditions used.

10.25 The following functional groups are described by the following Lewis structures:

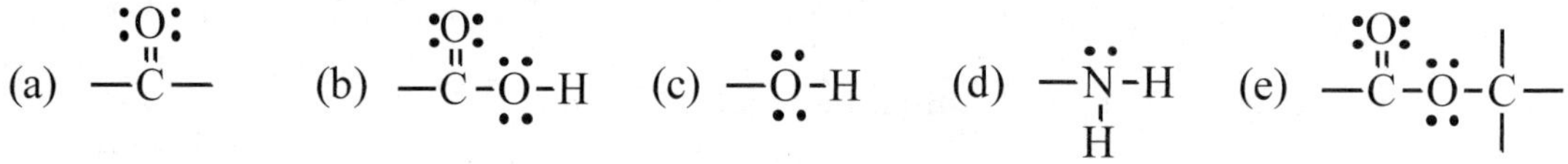

$$\text{(a)}\ \begin{array}{c}:\!\ddot{O}\!: \\ \| \\ -C- \end{array} \qquad \text{(b)}\ \begin{array}{c}:\!\ddot{O}\!: \\ \| \\ -C-\ddot{\underset{\cdot\cdot}{O}}-H \end{array} \qquad \text{(c)}\ -\ddot{\underset{\cdot\cdot}{O}}-H \qquad \text{(d)}\ \begin{array}{c} -\ddot{N}-H \\ | \\ H \end{array} \qquad \text{(e)}\ \begin{array}{c}:\!\ddot{O}\!:\quad\ | \\ \|\qquad\ \ \\ -C-\ddot{\underset{\cdot\cdot}{O}}-C- \\ \qquad\quad\ | \end{array}$$

10.27 (a) Incorrect. The carbon on the left has five bonds.
(b) Incorrect. The carbon bearing the Cl has five bonds.
(c) Correct.
(d) Incorrect. Oxygen has three bonds and one carbon has five bonds.
(e) Correct.
(f) Incorrect. Carbon on the right has five bonds.

10.29 The one tertiary alcohol with the molecular formula $C_4H_{10}O$ is:

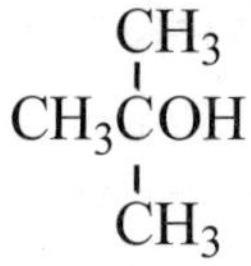

$$\begin{array}{c} CH_3 \\ | \\ CH_3COH \\ | \\ CH_3 \end{array}$$

10.31 The one tertiary amine with the molecular formula $C_4H_{11}N$ is:

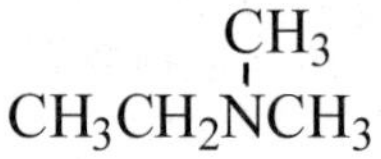

$$\begin{array}{c} \qquad CH_3 \\ \qquad | \\ CH_3CH_2NCH_3 \end{array}$$

10.33 (a) The four primary alcohols with the molecular formula $C_5H_{12}O$ are:

$$CH_3CH_2CH_2CH_2CH_2OH \quad \begin{array}{c} CH_3CHCH_2CH_2OH \\ | \\ CH_3 \end{array} \quad \begin{array}{c} CH_3CH_2CHCH_2OH \\ | \\ CH_3 \end{array} \quad \begin{array}{c} CH_3 \\ | \\ CH_3CCH_2OH \\ | \\ CH_3 \end{array}$$

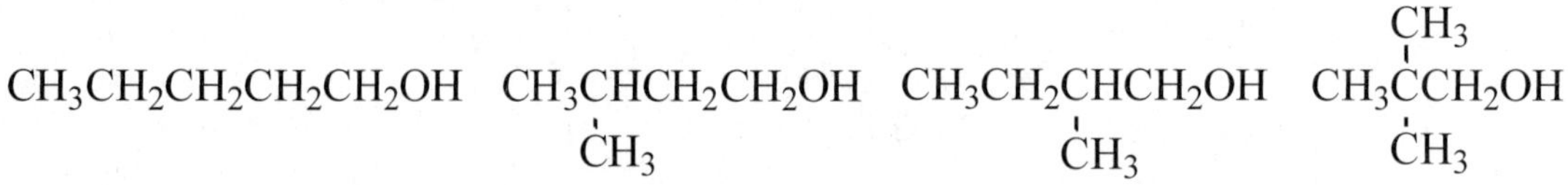

(b) The three secondary alcohols with the molecular formula $C_5H_{12}O$ are:

$$\begin{array}{c} OH \\ | \\ CH_3CHCH_2CH_2CH_3 \end{array} \qquad \begin{array}{c} OH \\ | \\ CH_3CH_2CHCH_2CH_3 \end{array} \qquad \begin{array}{c} OH \\ | \\ CH_3CHCHCH_3 \\ | \\ CH_3 \end{array}$$

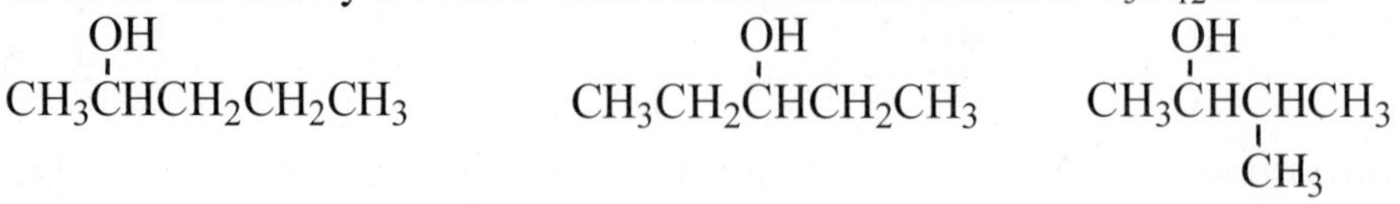

(c) The one tertiary alcohol with the molecular formula $C_5H_{12}O$ is:

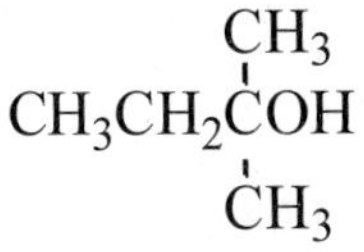

$$\begin{array}{r} CH_3 \\ | \\ CH_3CH_2COH \\ | \\ CH_3 \end{array}$$

<u>10.35</u> The eight carboxylic acids with the molecular formula $C_6H_{12}O_2$ are:

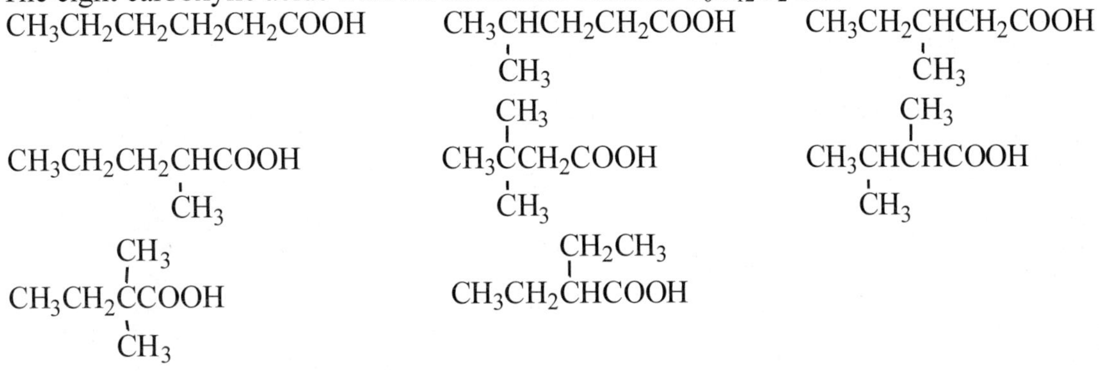

$$CH_3CH_2CH_2CH_2CH_2COOH$$

$$\begin{array}{l} CH_3CHCH_2CH_2COOH \\ \quad | \\ \quad CH_3 \end{array}$$

$$\begin{array}{l} CH_3CH_2CHCH_2COOH \\ \qquad\quad | \\ \qquad\quad CH_3 \end{array}$$

$$\begin{array}{l} CH_3CH_2CH_2CHCOOH \\ \qquad\qquad\quad | \\ \qquad\qquad\quad CH_3 \end{array}$$

$$\begin{array}{l} \quad CH_3 \\ \quad | \\ CH_3CCH_2COOH \\ \quad | \\ \quad CH_3 \end{array}$$

$$\begin{array}{l} \quad CH_3 \\ \quad | \\ CH_3CHCHCOOH \\ \quad | \\ \quad CH_3 \end{array}$$

$$\begin{array}{l} \qquad\quad CH_3 \\ \qquad\quad | \\ CH_3CH_2CCOOH \\ \qquad\quad | \\ \qquad\quad CH_3 \end{array}$$

$$\begin{array}{l} \qquad\quad CH_2CH_3 \\ \qquad\quad | \\ CH_3CH_2CHCOOH \end{array}$$

<u>10.37</u> Taxol was discovered by a survey of indigenous plants sponsored by the National Cancer Institute with the goal of discovering new chemicals for fighting cancer.

<u>10.39</u> The goal of combinatorial chemistry is to synthesize large numbers of closely related compounds in one reaction mixture.

<u>10.41</u> Predict 109.5° for each C-Si-C bond angle.

<u>10.43</u> The structures for the indicated chemical formula:

(a) CH_3CH_2OH (b) $\begin{array}{r} O \\ \| \\ CH_3CH_2CH \end{array}$ (c) $\begin{array}{c} \quad O \\ \quad \| \\ CH_3CCH_3 \end{array}$ (d) $\begin{array}{r} O\ \ \\ \|\ \ \\ CH_3CH_2COH \end{array}$

<u>10.45</u> The three tertiary amines with the molecular formula $C_5H_{13}N$ are:

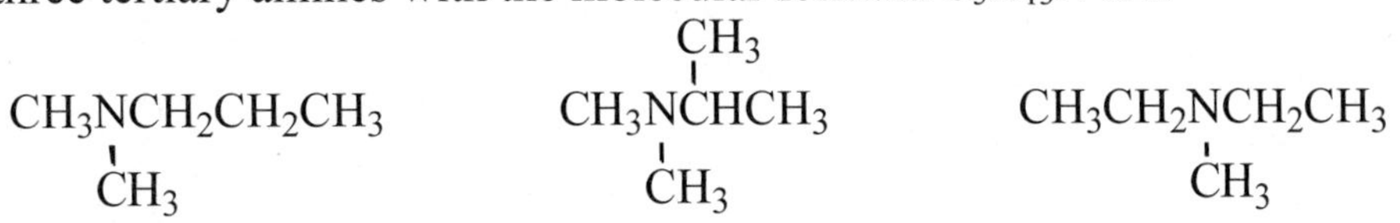

$$\begin{array}{l} CH_3NCH_2CH_2CH_3 \\ \quad | \\ \quad CH_3 \end{array}$$

$$\begin{array}{l} \qquad\quad CH_3 \\ \qquad\quad | \\ CH_3NCHCH_3 \\ \qquad | \\ \qquad CH_3 \end{array}$$

$$\begin{array}{l} CH_3CH_2NCH_2CH_3 \\ \qquad\quad | \\ \qquad\quad CH_3 \end{array}$$

<u>10.47</u> (a) The O-H is the most polar bond. (b) The C-C and C=C are the least polar bonds.

<u>10.49</u> (a) A secondary alcohol and a carboxylic acid
(b) Two primary alcohols
(c) One primary amine and one carboxylic acid
(d) One primary alcohol, one secondary alcohol, and one aldehyde
(e) Two primary amines

10.51 The following compounds have a molecular formula of $C_4H_8O_2$:

(a) Two carboxylic acids:

$$CH_3CH_2CH_2\overset{O}{\overset{\|}{C}}OH \qquad CH_3\underset{\underset{CH_3}{|}}{CH}\overset{O}{\overset{\|}{C}}OH$$

(b) Four esters:

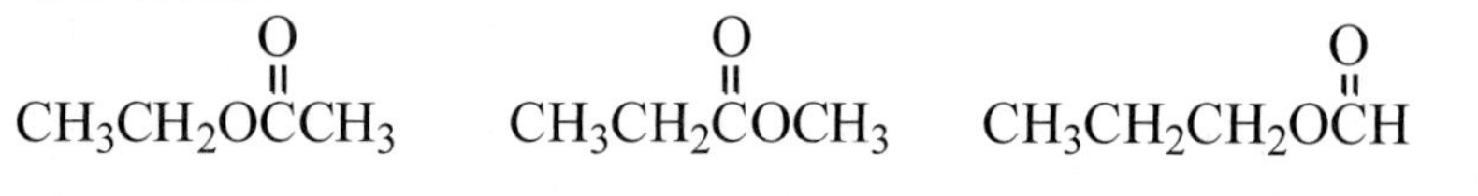

$$CH_3CH_2O\overset{O}{\overset{\|}{C}}CH_3 \qquad CH_3CH_2\overset{O}{\overset{\|}{C}}OCH_3 \qquad CH_3CH_2CH_2O\overset{O}{\overset{\|}{C}}H \qquad CH_3\underset{\underset{CH_3}{|}}{CH}O\overset{O}{\overset{\|}{C}}H$$

(c) One ketone with a 2° alcohol group:

$$CH_3\overset{O}{\overset{\|}{C}}\underset{\underset{OH}{|}}{CH}CH_3$$

(d) One aldehyde with a 3° alcohol group:

$$CH_3\overset{OH}{\underset{\underset{CH_3}{|}}{\overset{|}{C}}}-\overset{O}{\overset{\|}{C}}H$$

(e) One compound with a 1° alcohol group and a C=C:

$$HOCH_2CH{=}CHCH_2OH$$

11.1 The alkane is octane, and its molecular formula is C_8H_{18}.

11.3 The three constitutional isomers with the molecular formula C_5H_{12} are:

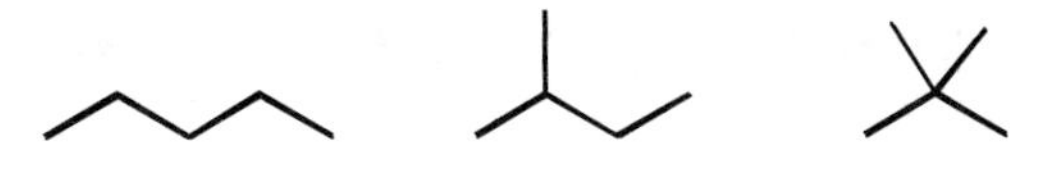

11.5 (a) Isobutylcyclopentane, C_9H_{18} (b) *sec-B*utylcycloheptane, $C_{11}H_{22}$
(c) 1-Ethyl-1-methylcyclopropane, C_6H_{12}

11.7 Cycloalkanes (a) and (c) show *cis-trans* isomerism.

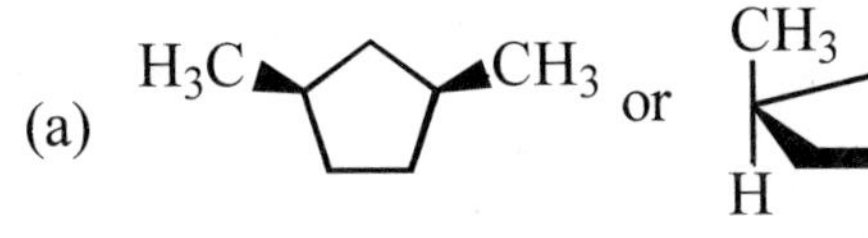

cis-1,3-Dimethylcyclopentane

trans-1,3-Dimethylcyclopentane

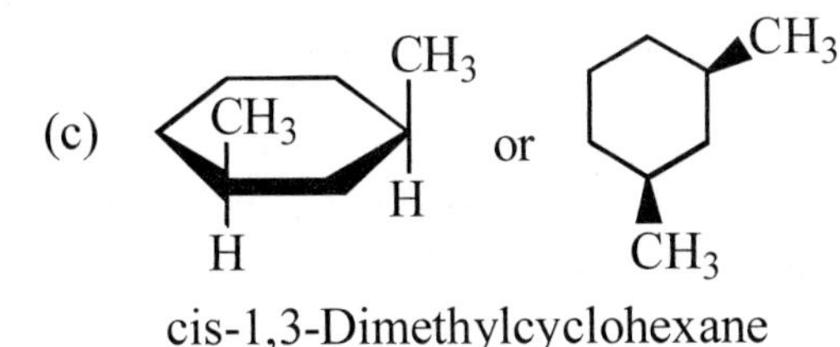

cis-1,3-Dimethylcyclohexane

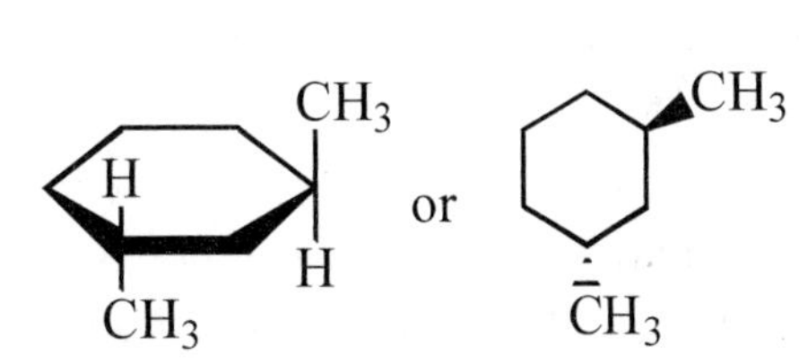

trans-1,3-Dimethylcyclohexane

11.9 The two products with the formula C_3H_7Cl are:

$CH_3CH_2CH_2Cl$
1-Chloropropane
(Propyl chloride)

$CH_3CH(Cl)CH_3$
2-Chloropropane
(Isopropyl chloride)

11.11 The carbon chain of an alkane is not straight; it is bent with all C-C-C bond angles of approximately 109.5°.

11.13 Line-angle formulas are:

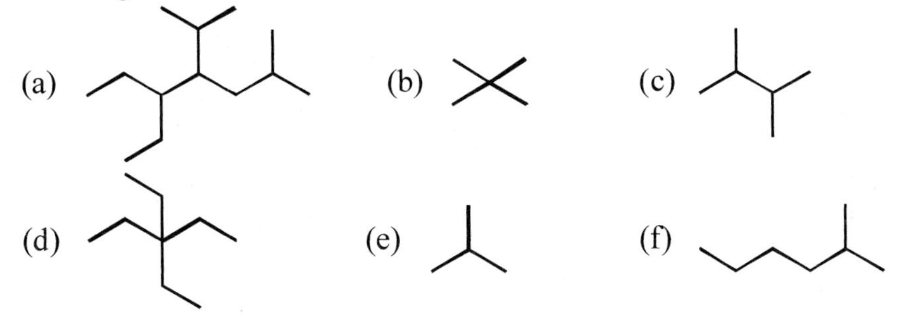

11.15 Statements (a) and (b) are true. Statements (c) and (d) are false.

11.17 Structures (a) and (g) represent the same compound. Structures (a,g), (c), (d), (e), and (f) represent constitutional isomers.

11.19 The structural formulas in parts (b), (c), (e), and (f) represent pairs of constitutional isomers.

11.21 (a) ethyl (b) isopropyl (c) isobutyl (d) *tert*-butyl

11.23 (a) 2-Methylpentane (b) 2,5-Dimethylhexane
(c) 3-Ethyloctane (d) 1-Iisopropyl-2-methylcyclohexane
(e) Isobutylcyclopentane (f) 1-Ethyl-2,4-dimethylcyclohexane

11.25 A conformation is any three-dimensional arrangement of the atoms in a molecule that results from rotation about a single bond.

11.27 Here are ball-and-stick models of two conformations of ethane. In the staggered conformation, the hydrogen atoms on one carbon are as far apart as possible from the hydrogens on the other carbon. In the eclipsed conformation, they are as close as possible.

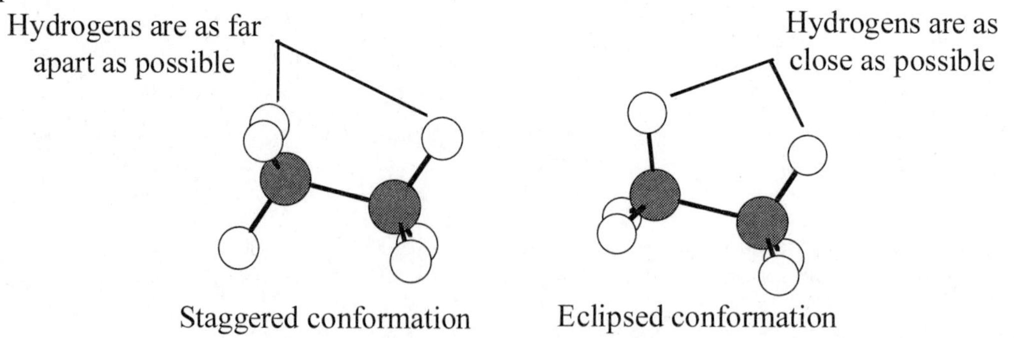

11.29 No. *Cis-trans* isomerism results when there is restricted rotation about carbon-carbon bonds. Alkanes have free rotation about their carbon-carbon bonds.

11.31 Structural formulas for the six cycloalkanes with the molecular formula C_5H_{10} are:

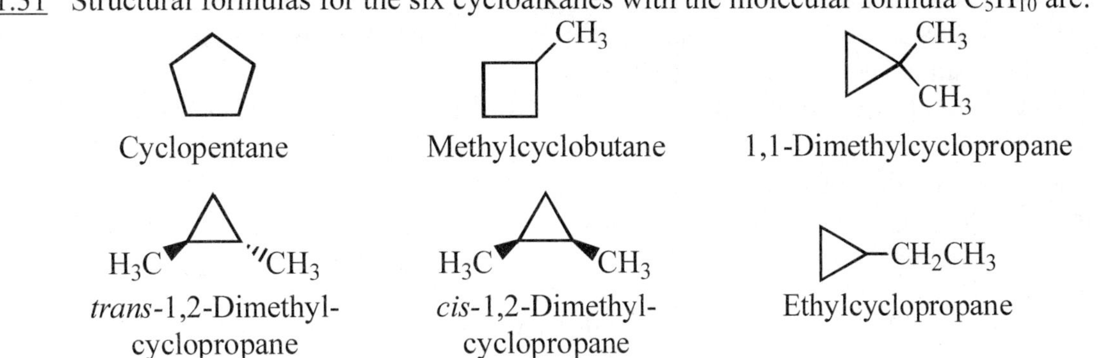

11.33 The isomer with the highest boiling point is heptane (98°C), which is unbranched. The isomer with the lowest boiling point is 2,2-dimethylpentane (79°C), one of the most branched isomers.

11.35 Alkanes are less dense than water. Those that are liquid at room temperature float on water.

11.37 Both hexane and octane are colorless liquids and you cannot tell the difference between them by looking at them. One way to tell which compound is which is to determine their boiling points. Hexane has the lower molecular weight, the smaller molecules, and the lower boiling point (69°C). Octane has the higher molecular weight, the larger molecules, and the higher boiling point (126°C).

11.39 The greater the molecular weight, the higher the boiling point.

11.41 On a gram-per-gram basis, methane is the better source of heat energy.

Hydrocarbon	Component of	Heat of combustion (kcal/mol)	Molar mass (g/mol)	Heat of combusion (kcal/g)
CH_4	natural gas	212	16.0	13.3
$CH_3CH_2CH_3$	LPG	530	44.1	12.0

11.43 The three chloropentanes are:

Cl

Cl

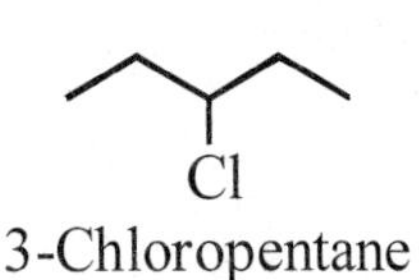

1-Chloropentane 2-Chloropentane 3-Chloropentane

11.45 (a) The Freons are a class of chlorofluorocarbons.
(b) They were ideal for use as heat transfer agents in refrigeration systems because they are nontoxic, not corrosive, and nonflammable.
(c) Two Freons used for this purpose were CCl_3F (Freon-11) and CCl_2F_2 (Freon-12).

11.47 They are hydrofluorocarbons and hydrochlorofluorocarbons. These compounds are much more chemically reactive in the atmosphere than the Freons and are destroyed before they reach the stratosphere.

11.49 Octane will produce more engine knocking than heptane.

11.51 (a) Constitutional isomers (b) Constitutional isomers
(c) Constitutional isomers (d) Not isomers
(e) Not isomers (f) Constitutional isomers

11.53 Compounds (a) and (c) show cis-trans isomerism.

(a) 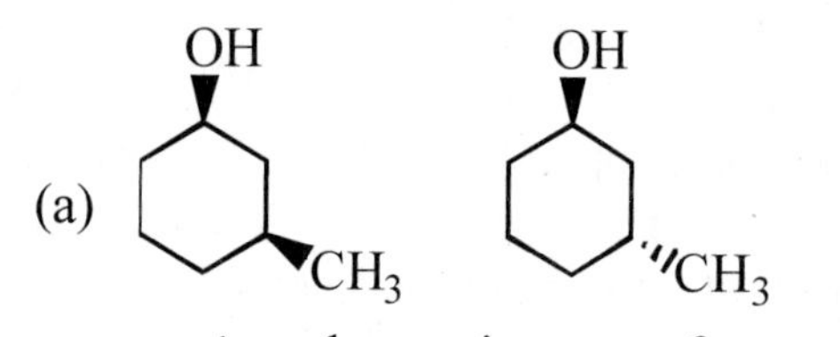

cis and *trans* isomers of 3-methylcyclohexanol

(c) 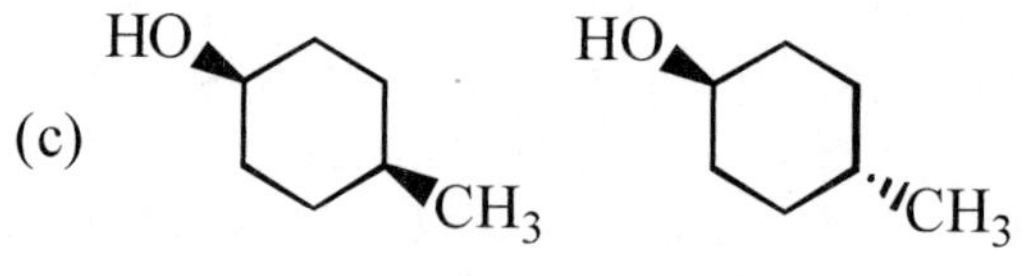

cis and *trans* isomers of 4-methylcyclohexanol

11.55 Dodecane (a) does not dissolve in water, (b) dissolves in hexane, (c) burns when ignited, (d) is a liquid at room temperature, and (e) is less dense that water.

11.57 In the following drawing, all hydrogens are shown.

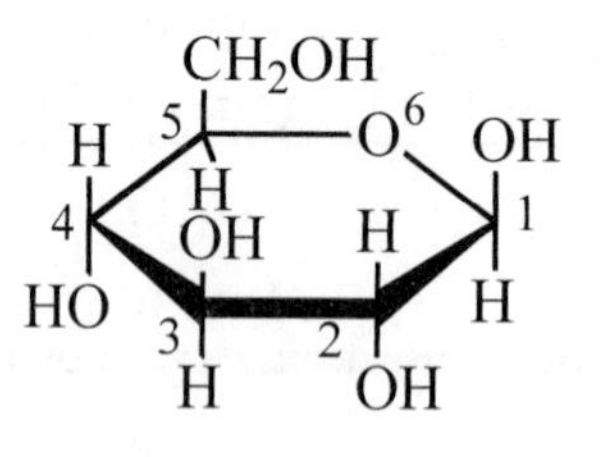

11.59 Water, a polar molecule, cannot penetrate the surface layer created by this nonpolar hydrocarbon.

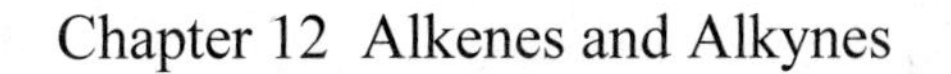

Chapter 12 Alkenes and Alkynes

12.1 (a) 3,3-Dimethyl-1-pentene (b) 2,3-Dimethyl-2-butene (c) 3,3-Dimethyl-1-butyne

12.3 (a) 1-Isopropyl-4-methylcyclohexene (b) Cyclooctene (c) 4-*tert*-Butylcyclohexene

12.5 Four *cis-trans* isomers are possible.

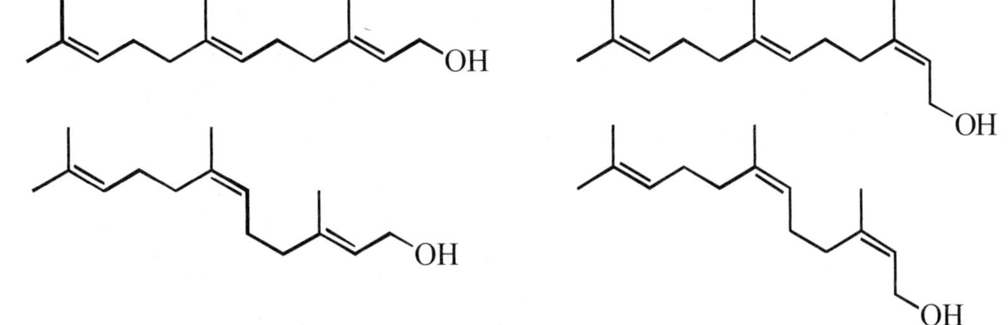

12.7 Propose a two-step mechanism similar to that for the addition of HCl to propene.

Step 1: Reaction of H^+ with the carbon-carbon double bond gives a 3° carbocation intermediate.

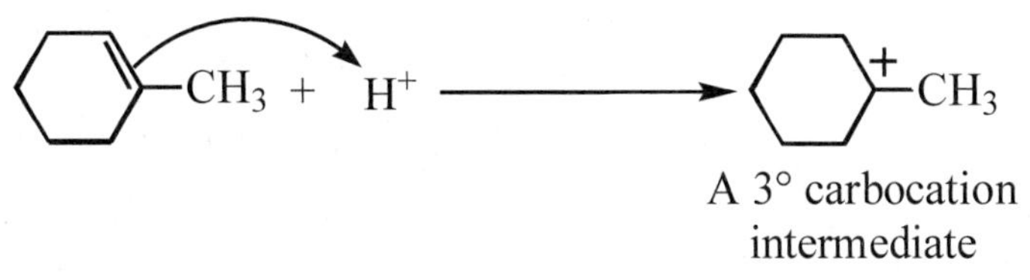

Step 2: Reaction of the 3° carbocation intermediate with bromide ion completes the valence shell of carbon and gives the product.

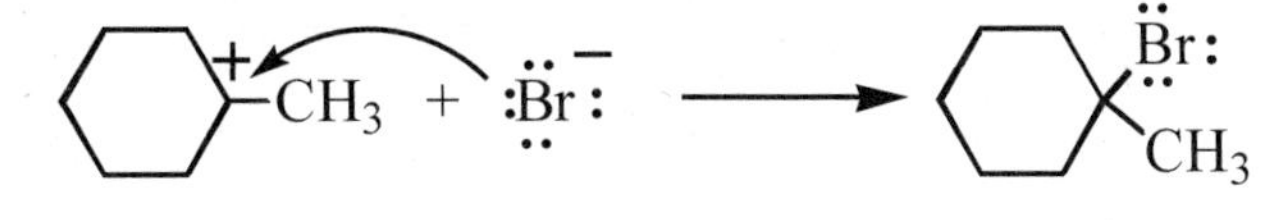

12.9 Propose a three-step mechanism similar to that for the acid-catalyzed hydration of propene.

Step 1: Reaction of the carbon-carbon double bond with H^+ gives a 3° carbocation intermediate.

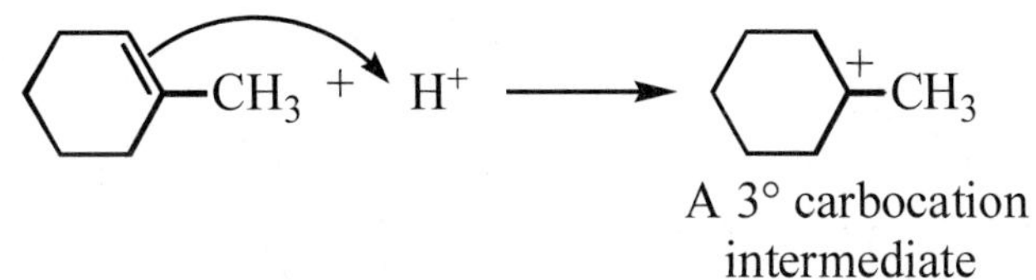

Step 2: Reaction of the 3° carbocation intermediate with water completes the valence shell of carbon and gives an oxonium ion.

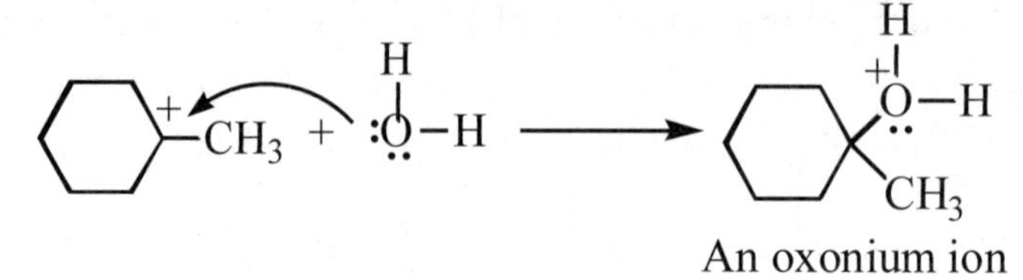

Step 3: Loss of H^+ from the oxonium ion completes the reaction and generates a new H^+ catalyst.

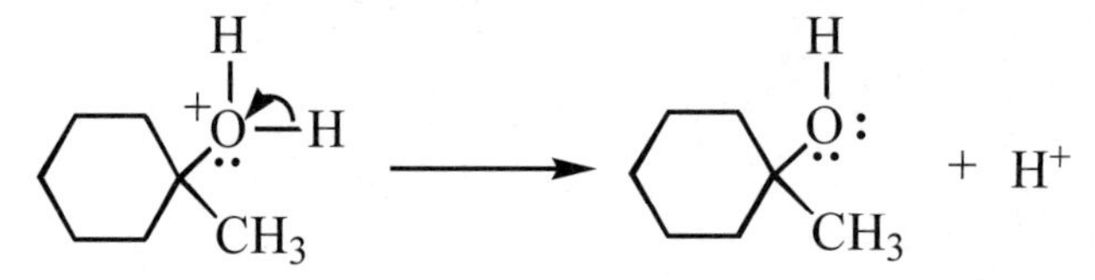

12.11 A saturated hydrocarbon contains only carbon-carbon single bonds. An unsaturated hydrocarbon contains one or more carbon-carbon double or triple bonds.

12.13

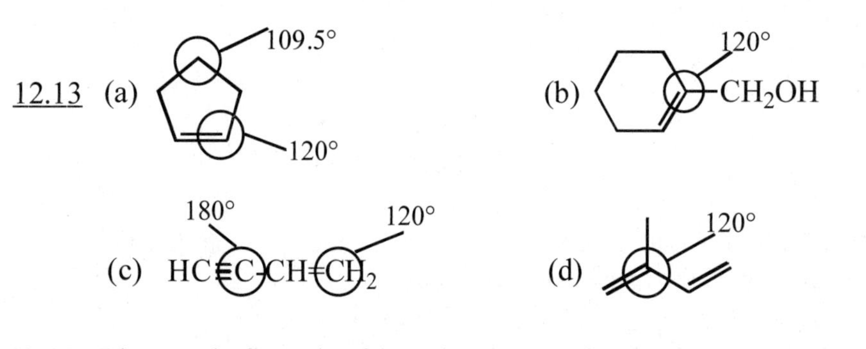

12.15 Line-angle formulas for each compound are

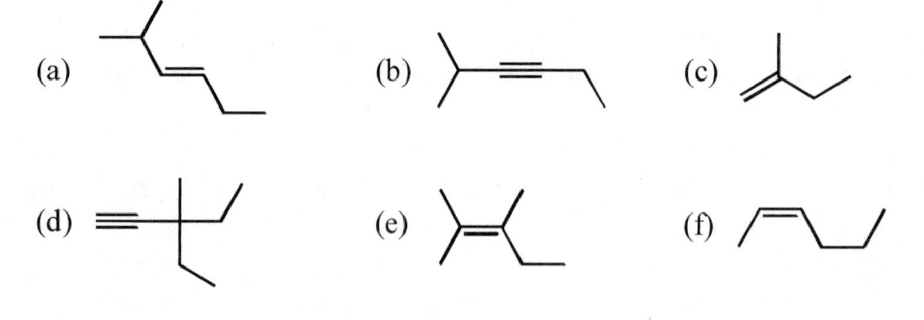

12.17 (a) 1-Heptene
(b) 1,4,4-Trimethylcyclopentene
(c) 1,3-Dimethylcyclohexene
(d) 2,4-Dimethyl-2-pentene
(e) 1-Octyne
(f) 2,2-Dimethyl-3-hexyne

12.19 (a) The longest chain is four carbons. The correct name is 2-butene.
(b) The chain is not numbered correctly. The correct name is 2-pentene.
(c) The ring is not numbered correctly. The correct name is 1-methylcyclohexene.
(d) The double bond must be located. The correct name is 3,3-dimethyl-1-pentene.
(e) The chain is not numbered correctly. The correct name is 2-hexyne.
(f) The longest chain is five carbon atoms. The correct name is 3,4-dimethyl-2-pentene.

12.21 The structural feature that makes *cis-trans* isomerism possible in alkenes is restricted rotation about the carbon-carbon double bond. In cycloalkanes, it is restricted rotation about the carbon-carbon bonds of the ring. These two structural features have in common restricted rotation about carbon-carbon bonds.

12.23 Only part (b), 2-pentene, shows cis-trans isomerism.

cis-2-Pentene *trans*-2-Pentene

12.25 The line-angle drawing of the all *cis* arachidonic acid:

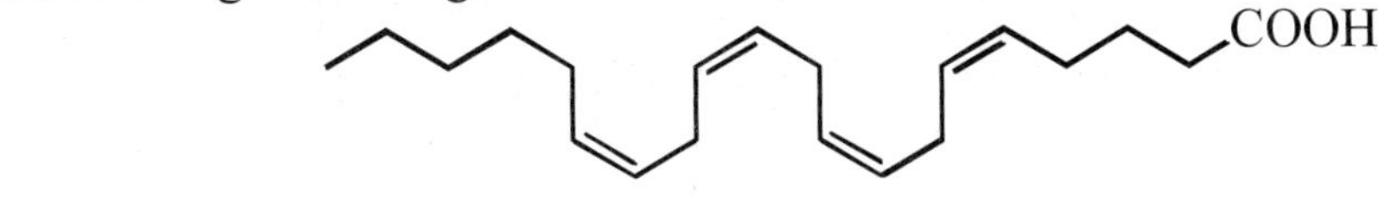

Arachidonic acid

12.27 Only compounds (b) and (d) show cis-trans isomerism.

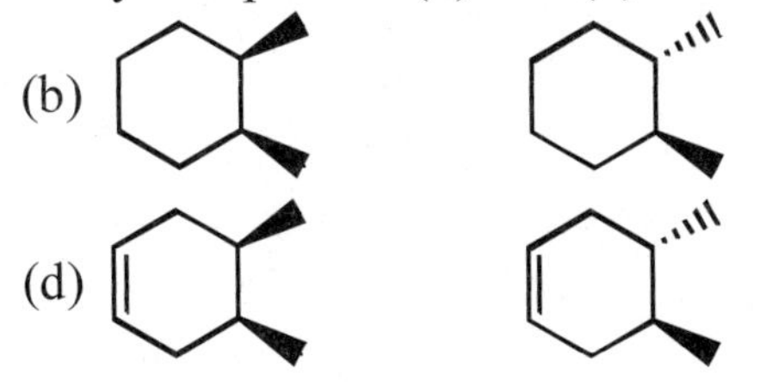

12.29 The structural formula of β-ocimene is drawn on the left showing all atoms and on the right as a line-angle formula.

$$CH_2{=}CH{-}C(CH_3){=}CH{-}CH_2CH{=}C(CH_3)CH_3$$

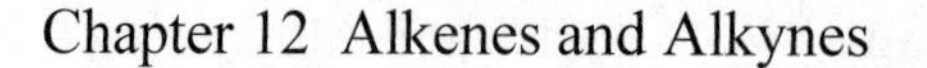

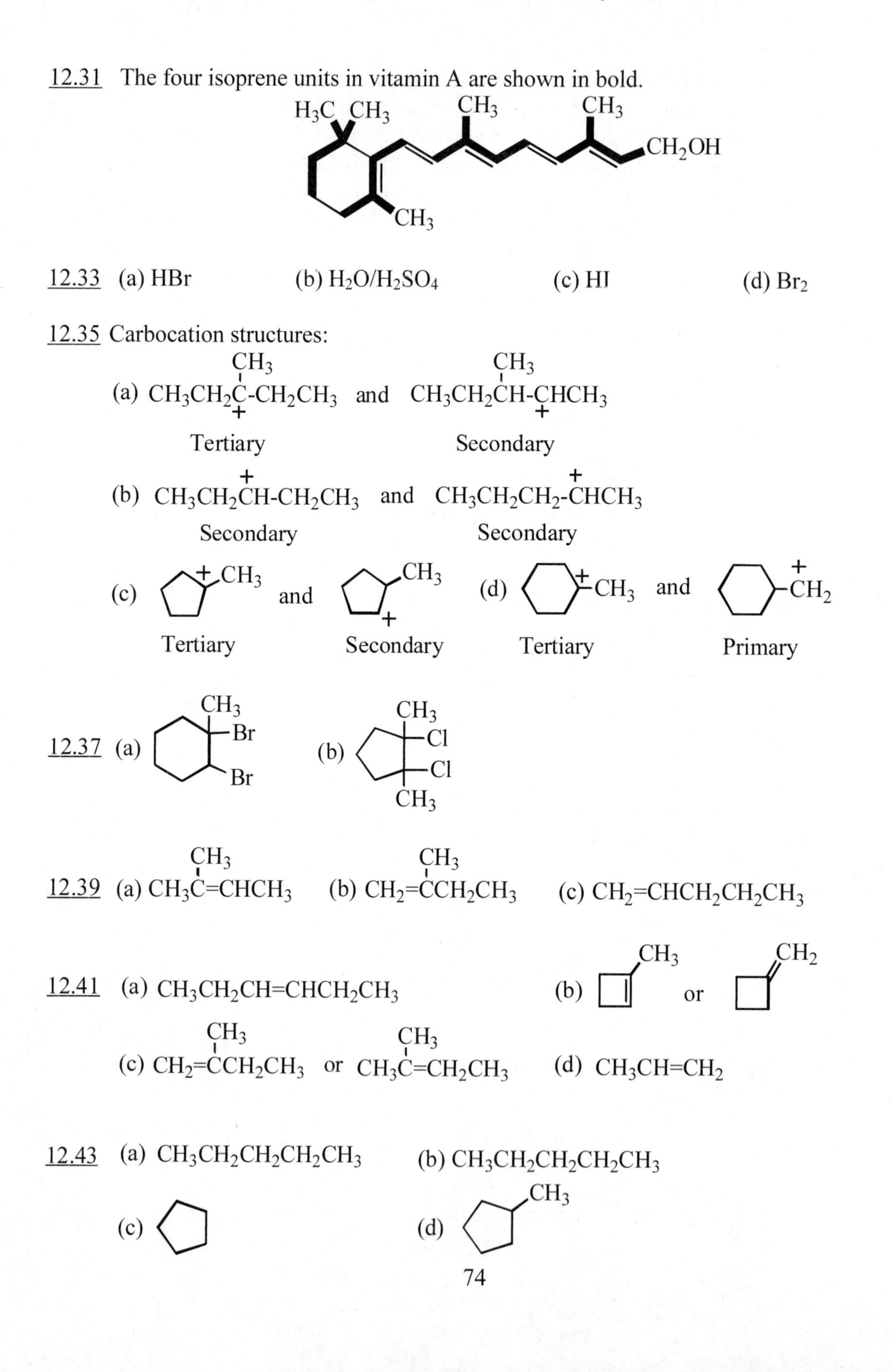

12.31 The four isoprene units in vitamin A are shown in bold.

12.33 (a) HBr (b) H_2O/H_2SO_4 (c) HI (d) Br_2

12.35 Carbocation structures:

(a) $CH_3CH_2\overset{+}{C}(CH_3)$-$CH_2CH_3$ and $CH_3CH_2CH(CH_3)$-$\overset{+}{C}HCH_3$

Tertiary Secondary

(b) $CH_3CH_2\overset{+}{C}H$-CH_2CH_3 and $CH_3CH_2CH_2$-$\overset{+}{C}HCH_3$

Secondary Secondary

(c) Tertiary and Secondary (d) Tertiary and Primary

12.37 (a) (b)

12.39 (a) $CH_3C(CH_3)$=$CHCH_3$ (b) CH_2=$C(CH_3)CH_2CH_3$ (c) CH_2=$CHCH_2CH_2CH_3$

12.41 (a) CH_3CH_2CH=$CHCH_2CH_3$ (b) or

(c) CH_2=$C(CH_3)CH_2CH_3$ or $CH_3C(CH_3)$=CH_2CH_3 (d) CH_3CH=CH_2

12.43 (a) $CH_3CH_2CH_2CH_2CH_3$ (b) $CH_3CH_2CH_2CH_2CH_3$

(c) (d)

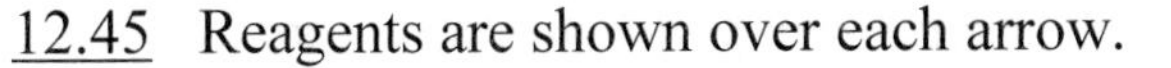

12.45 Reagents are shown over each arrow.

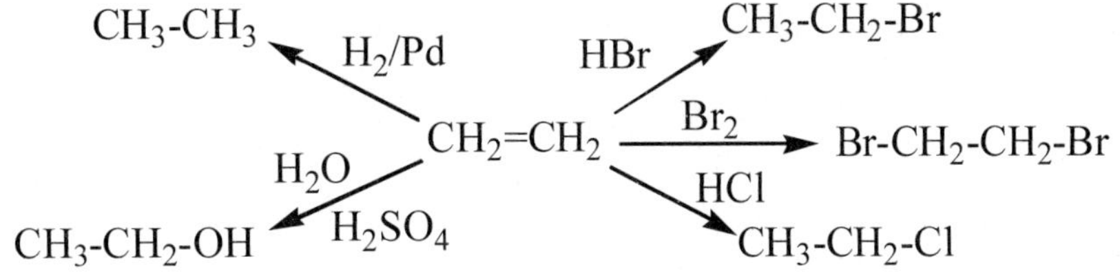

12.47 Ethylene is a natural ripening agent for fruits.

12.49 Its molecular formula is $C_{16}H_{30}O_2$ and its molecular weight is 254.4 amu, and molar mass is 254.4 g/mol.

12.51 Rods are primarily responsible for black-and-white vision. Cones are responsible for color vision.

12.53 The most common consumer items made of high-density polyethylene (HDPE) are milk and water jugs, grocery bags, and squeeze bottles. The most common consumer items made of low-density polyethylene (LDPE) are packaging for baked goods, vegetables and other produce, as well as trash bags. Currently only HDPE materials are recyclable.

12.55 There are five compounds with the molecular formula C_4H_8. All are constitutional isomers. The only *cis-trans* isomers are *cis*-2-butene and *trans*-2-butene.

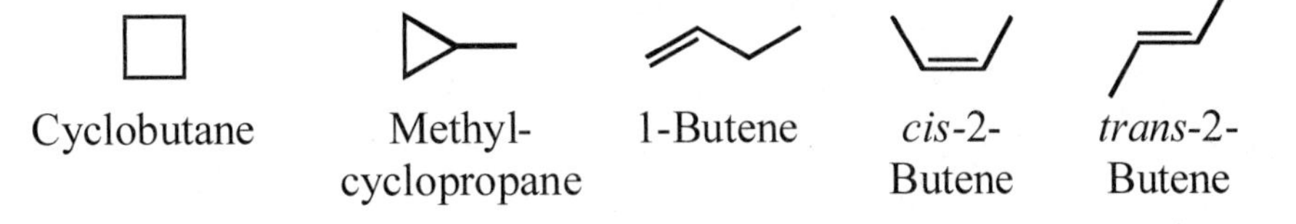

12.57 (a) The carbon skeleton of lycopene can be divided into eight isoprene units, here shown in bold bonds.

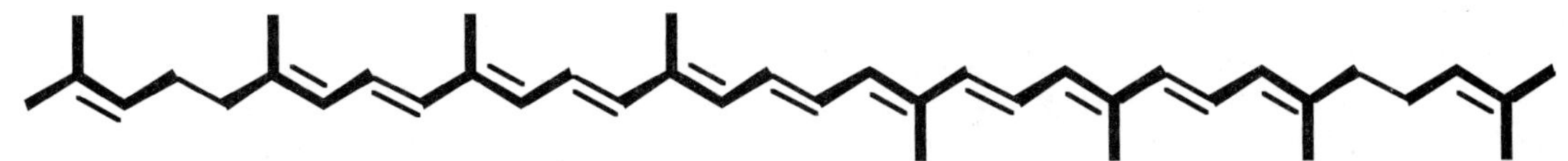

(b) Eleven of the 13 double bonds have the possibility for cis-trans isomerism. The double bonds at either end of the molecule cannot show cis-trans isomerism.

12.59

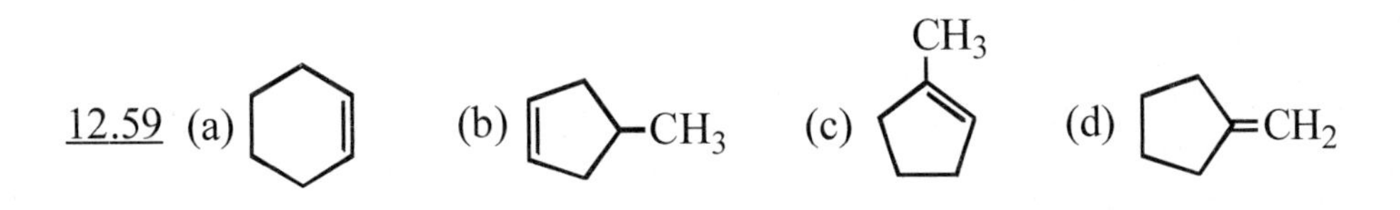

12.61 The alcohol is 3-hexanol. Each alkene gives the same 2° carbocation intermediate and the same alcohol.

12.63 Reagents are shown over the arrows.

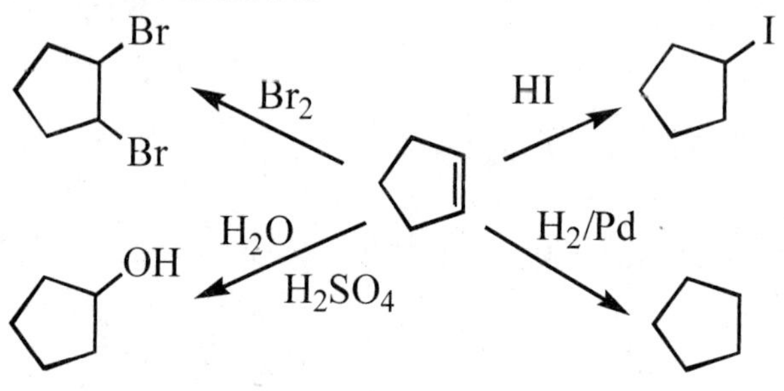

12.65 Two *cis-trans* isomers are possible for oleic acid, four for linoleic acid, and eight for linolenic acid.

13.1 (a) 2,4,6-Tri-*tert*-butylphenol (b) 2,4-Dichloroaniline
(c) 3-Nitrobenzoic acid

13.3 An aromatic compound is one that contains one or more benzene rings.

13.5 Yes, they have double bonds, at least in the contributing structures we normally use to represent them. Yes, they are unsaturated because they have fewer hydrogens than a cycloalkane with the same number of carbons.

13.7 (a) An alkene of six carbons has the molecular formula C_6H_{12} and contains one carbon-carbon double bond. Three examples are:

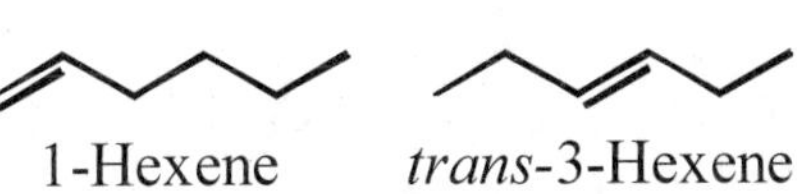

1-Hexene *trans*-3-Hexene *cis*-3-Hexene

(b) A cycloalkene of six carbons has the molecular formula C_6H_{10} and contains one ring and one carbon-carbon double bond. Three examples are:

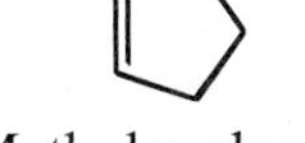

Cyclohexene 4-Methylcyclopentene 1-Methylcyclopentene

(c) An alkyne of six carbons has the molecular formula C_6H_{10} and contains one carbon-carbon triple bond. Three examples are:

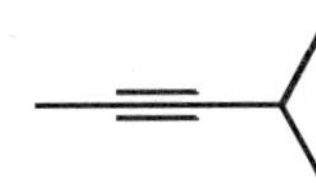

1-Hexyne 2-Hexyne 4-Methyl-2-pentyne

(d) An aromatic hydrocarbon of eight carbons has the molecular formula C_8H_{10} and contains one benzene ring. Three examples are:

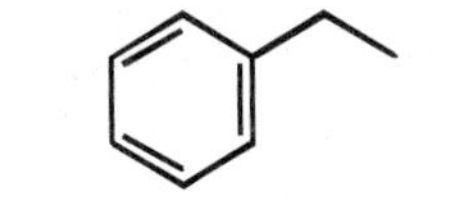

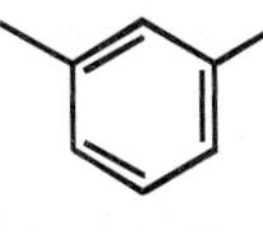

Ethylbenzene 1,3-Dimethylbenzene (*m*-Xylene) 1,4-Dimethylbenzene (*p*-Xylene)

13.9 Benzene consists of carbons, each surrounded by three regions of electron density, which gives 120° for all bond angles. Bond angles of 120° in benzene can be maintained only if the molecule is planar. Cyclohexane, on the other hand, consists of carbons, each surrounded by four regions of electron density, which gives 109.5° for all bond angles. Angles of 109.5° in cyclohexane can be maintained only if the molecule is nonplanar.

13.11 It works this way. Neither a unicorn, which has a horn like a rhinoceros, nor a dragon, which has a tough, leathery hide like a rhinoceros, exists. If they did and you made a hybrid of them, you would have a rhinoceros. Furthermore, a rhinoceros is not a dragon part of the time and a unicorn the rest of the time; a rhinoceros is a rhinoceros all of the time. To carry this analogy to aromatic compounds, resonance-contributing structures for them do not exist; they are imaginary. If they did exist and you could make a hybrid of them, you would have the real aromatic compound.

13.13

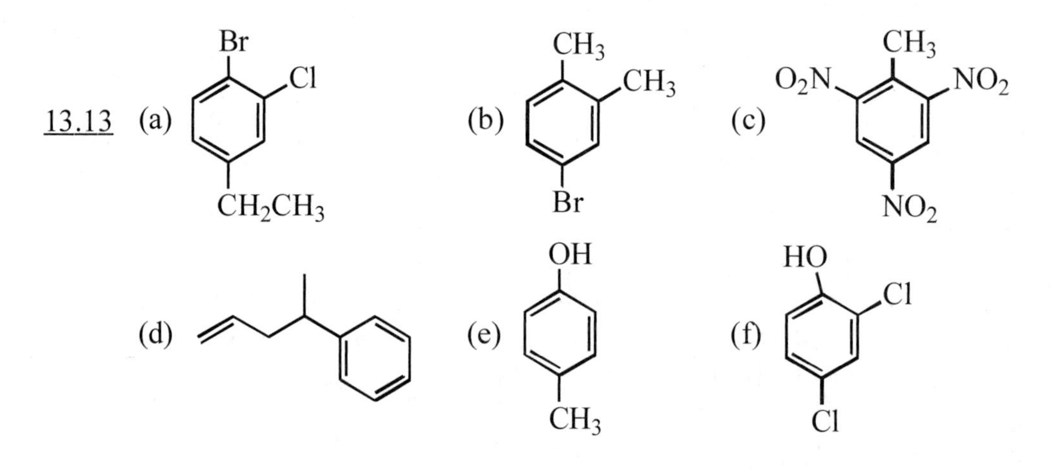

13.15 Only cyclohexene will react with a solution of bromine in dichloromethane. A solution of Br_2/CH_2Cl_2 is red, whereas a dibromocycloalkane is colorless. To distinguish which bottle contains which compound, place a small quantity of each compound in a test tube and to each add a few drops of Br_2/CH_2Cl_2 solution. If the red color disappears, the compound is cyclohexene. If it remains, the compound is benzene.

13.17

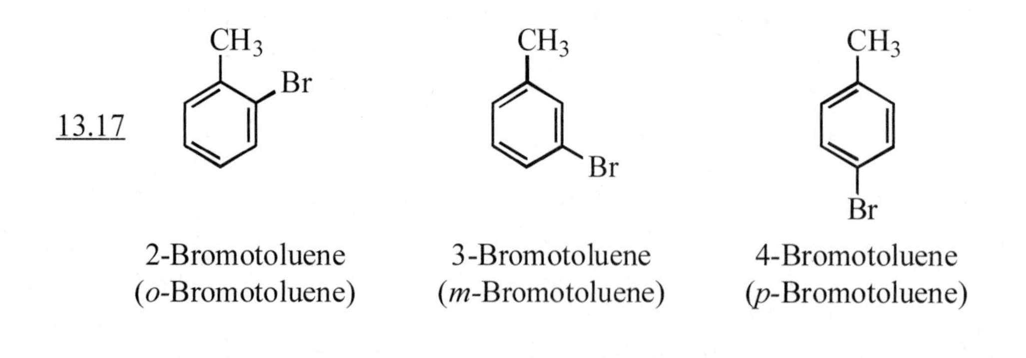

2-Bromotoluene (*o*-Bromotoluene) 3-Bromotoluene (*m*-Bromotoluene) 4-Bromotoluene (*p*-Bromotoluene)

13.19 (a) Nitration using HNO_3/H_2SO_4 followed by sulfonation using H_2SO_4. The order of the steps may also be reversed.
(b) Bromination using $Br_2/FeCl_3$ followed by chlorination using $Cl_2/FeCl_3$. The order of the steps may also be reversed.

13.21 Phenol is a sufficiently strong acid that it reacts with strong bases such as sodium hydroxide to form sodium phenoxide, a water-soluble salt. Cyclohexanol has no comparable acidity and does not react with sodium hydroxide.

13.23 *Radical* indicates a molecule or ion with an unpaired electron. *Chain* means a cycle of two or more steps, called propagation steps that repeat over and over. The net effect of a radical chain reaction is the conversion of starting materials to products. *Chain length* refers to the number of times the cycle of chain propagation steps repeats.

13.25 Vitamin E participates in one or the other of the chain propagation steps and forms a stable radical, which breaks the cycle of propagation steps.

13.27 The abbreviation DDT is derived from **D**ichloro**D**iphenyl**T**richloroethane.

13.29 DDT is expected to be insoluble in water. There are no hydrophilic functional groups such as –OH or –COOH that would hydrogen bond with water. It is entirely a hydrophobic molecule.

13.31 Biodegradable means that a substance can be broken down into one or more harmless compounds by living organisms in the environment.

13.33 Iodine is an element that is found primarily in seawater and, therefore, seafood is a rich source of it. Individuals in inland areas where seafood is only a limited part of the diet are the most susceptible to developing goiter.

13.35 One of the aromatic rings in Allura red has a -CH_3 group and an -OCH_3 group. These groups are not present in Sunset Yellow.

13.37 Orange

13.39 Capsaicin is isolated from the fruit of various species of *Capsicum*, otherwise known as chili peppers.

13.41 Two *cis-trans* isomers are possible for capsaicin. Capsaicin exists as the *trans* stereoisomer.

13.43 Following are the three contributing structures for naphthalene.

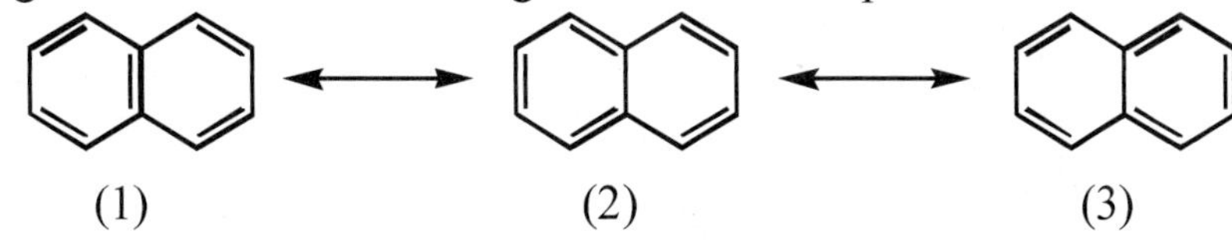

13.45 BHT participates in one of the chain propagation steps of autoxidation, forming a stable radical and thus terminating autoxidation.

13.47

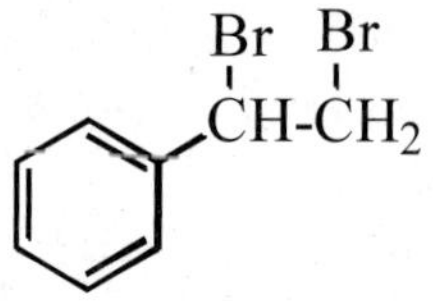

Br Br

$CH\text{-}CH_2$

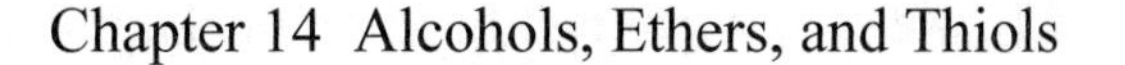

Chapter 14 Alcohols, Ethers, and Thiols

14.1 (a) 2-Heptanol (b) 2,2-Dimethyl-1-propanol (c) *cis*-3-Isopropylcyclohexanol

14.3 In each case, the major product (circled) contains the more substituted double bond.

(a) $CH_3C(CH_3)=CHCH_3$ + $CH_2=C(CH_3)CH_2CH_3$ (b) 1-methylcyclopentene (CH_3) + methylenecyclopentane ($=CH_2$)

14.5 Each secondary alcohol is oxidized to a ketone.

(a) cyclohexanone (=O) (b) $CH_3C(=O)CH_2CH_2CH_3$

14.7 (a) 3-Methyl-1-butanethiol (b) 3-Methyl-2-butanethiol

14.9 Only (c) and (d) are secondary alcohols.

14.11 (a) 1-Pentanol (b) 1,3-Propanediol
(c) 1,2-Butanediol (d) 3-Methyl-1-butanol
(e) *cis*-1,2-Cyclohexanediol (f) 2,6-Dimethylcyclohexanol

14.13

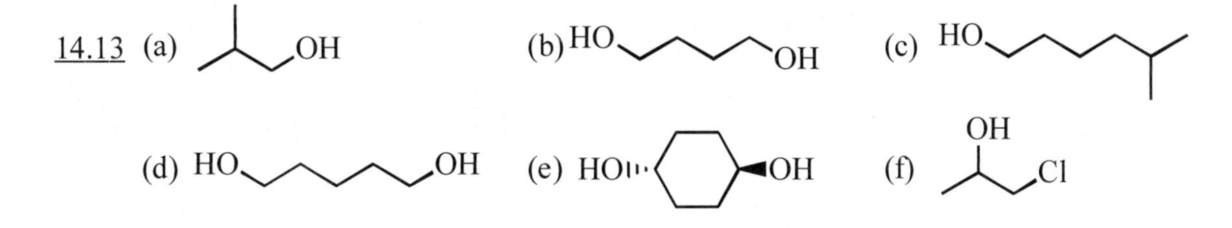

14.15 (a) Prednisone contains three ketones, one primary alcohol, one tertiary alcohol, one disubstituted carbon-carbon double bond, and one trisubstituted carbon-carbon double bond.
(b) Estradiol contains one secondary alcohol and one disubstituted phenol.

14.17 Low-molecular-weight alcohols form hydrogen bonds with water molecules through both the oxygen and hydrogen atoms of their –OH groups. Low-molecular-weight ethers form hydrogen bonds with water molecules only through the oxygen atom of their –O– groups. The greater extent of hydrogen bonding between alcohol and water molecules makes the low-molecular-weight alcohols more soluble in water than the low-molecular-weight ethers.

14.19 The following illustration describes (a) the hydrogen bonding between the oxygen of methanol and the hydrogen of water; and (b) the hydrogen bonding between the hydrogen of methanol's –OH group and the oxygen of water.

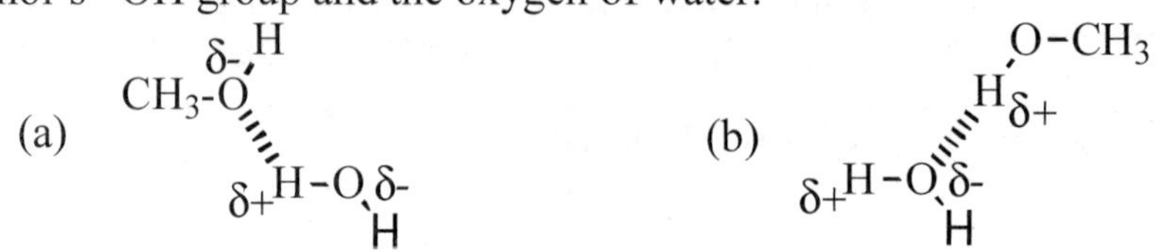

14.21 In order of increasing boiling point, they are:

$CH_3CH_2CH_3$	CH_3CH_2OH	$CH_3CH_2CH_2CH_2OH$	$HOCH_2CH_2OH$
-42°C	78°C	117°C	198°C

14.23 Evaporation of a liquid from the surface of the skin cools because heat is absorbed from the skin in converting molecules from the liquid state to the gaseous state. 2-Propanol (isopropyl alcohol), which has a boiling point of 82°C, absorbs heat from the skin, evaporates rapidly, and has a cooling effect. 2-Hexanol, which has a boiling point of 140°C, also absorbs heat from the surface of the skin but, because of its higher boiling point, evaporates much more slowly and, therefore, does not have the same cooling effect as 2-propanol.

14.25 The more water-soluble compound is circled.

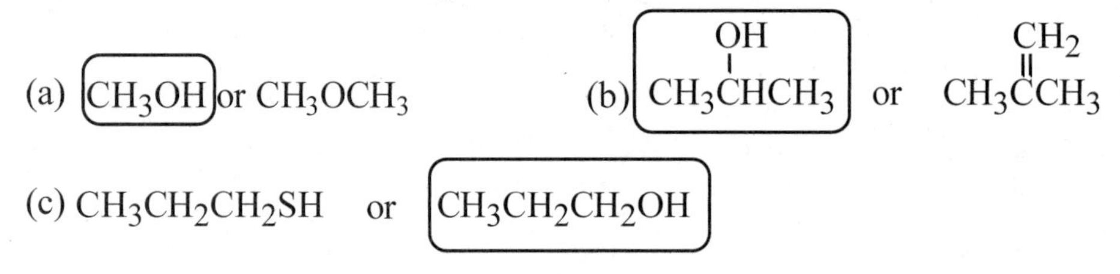

14.27 For three parts, two constitutional isomers will give the desired alcohol. For two parts, only one alkene will give the desired alcohol.

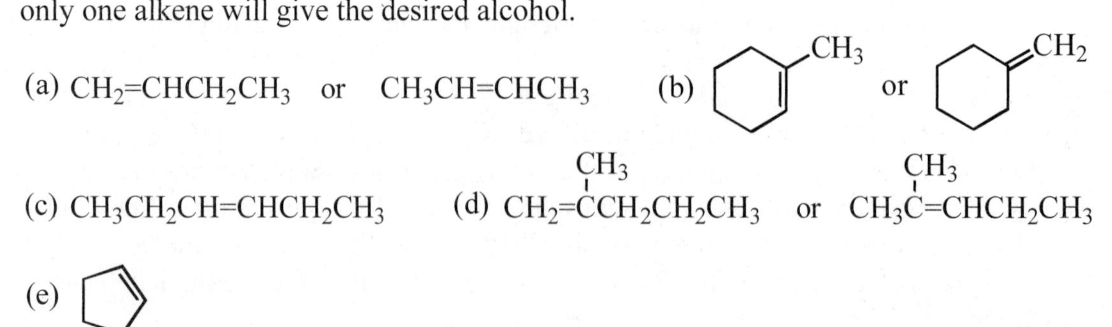

14.29 Phenols are weak acids, with pK_a values approximately equal to 10. Alcohols, considerably weaker acids, have about the same acidity as water.

14.31 The first reaction is an acid-catalyzed dehydration; the second is an oxidation.

(a) $CH_3CH_2CH_2CH_2OH \xrightarrow[heat]{H_2SO_4} CH_3CH_2CH{=}CH_2 + H_2O$

(b) $CH_3CH_2CH_2CH_2OH \xrightarrow[H_2SO_4]{K_2Cr_2O_7} CH_3CH_2CH_2\overset{O}{\overset{\|}{C}}OH$

14.33 Oxidation of a primary alcohol by $K_2Cr_2O_7/H_2SO_4$ gives a carboxylic acid.

(a) $CH_3(CH_2)_6\overset{O}{\overset{\|}{C}}OH$ (b) $HO\overset{O}{\overset{\|}{C}}CH_2CH_2\overset{O}{\overset{\|}{C}}OH$

14.35 Each can be prepared from 1-propanol (circled) as shown in this flow chart.

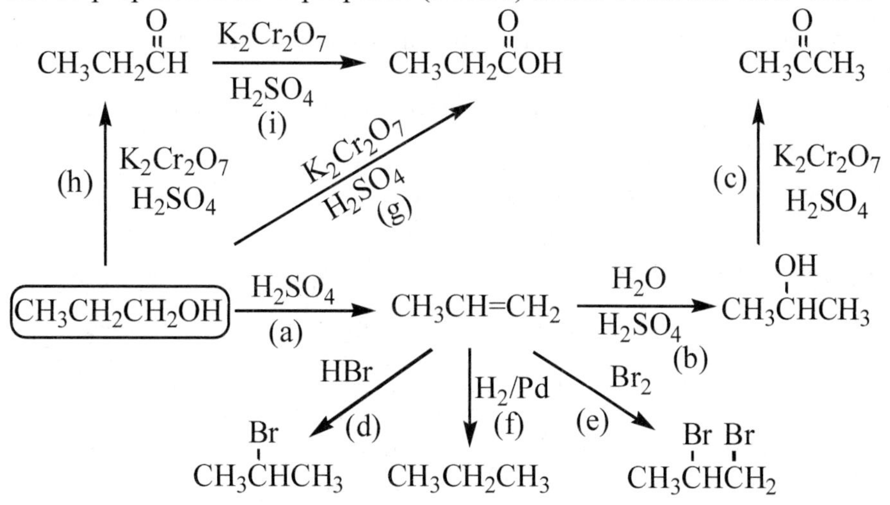

14.37 Ethanol and ethylene glycol are derived from ethylene. Ethanol is a solvent and is the starting material for the synthesis of diethyl ether, also an important solvent. Ethylene glycol is used in automotive antifreezes and is one of the two starting materials required for the synthesis of poly(ethylene terephthalate), better known as PET (Section 19.6B).

14.39 (a) Dicyclopentyl ether (b) Dipentyl ether (c) Diisopropyl ether

14.41 (a) 2-Butanethiol (b) 1-Butanethiol (c) Cyclohexanethiol

14.43 Because 1-butanol molecules associate in the liquid state by hydrogen bonding, it has the higher boiling point (117°C). There is very little polarity to an S-H bond. The only interactions among 1-butanethiol molecules in the liquid state are the considerably weaker London dispersion forces. For this reason, 1-butanethiol has the lower boiling point (98°C).

14.45 Nitroglycerin was discovered in 1847. It is a pale yellow, oily liquid.

14.47 One of the products the body derives from metabolism of nitroglycerin is nitric oxide, NO, which causes the coronary artery to dilate, thus relieving angina.

14.49 The relationship is that 2100 mL of breath contains the same amount of ethanol as 1.00 mL of blood.

14.51 Diethyl ether is easy to use and causes excellent muscle relaxation. Blood pressure, pulse rate, and respiration are usually only slightly affected. Diethyl ether's chief drawbacks are its irritating effect on the respiratory passages and its after-effect of nausea.

14.53 Enflurane and isoflurane are insoluble in water but soluble in hexane.

14.55 $2CH_3OH + 3O_2 \longrightarrow 2CO_2 + 4H_2O$

14.57 The eight alcohol constitutional isomers with the molecular formula $C_5H_{12}O$ are:

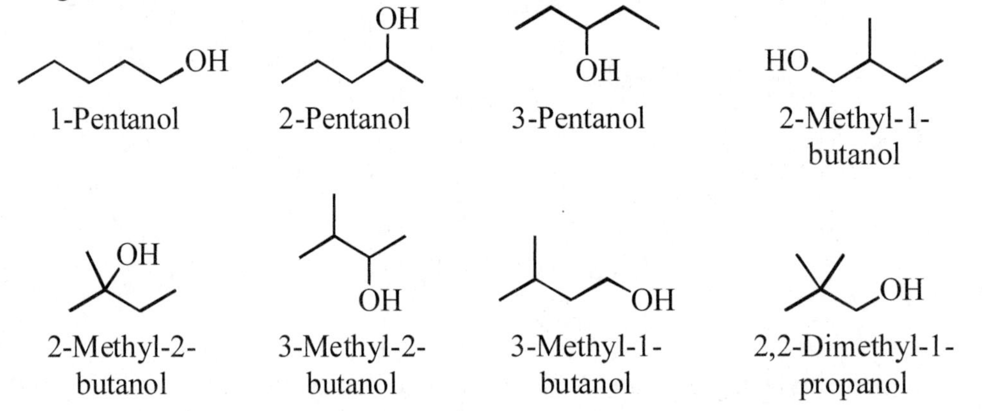

14.59 Ethylene glycol has two -OH groups by which each molecule participates in hydrogen bonding, whereas 1-propanol has only one. The stronger intermolecular forces of attraction between molecules of ethylene glycol give it the higher boiling point.

14.61 Arranged in order of increasing solubility in water, they are:

$CH_3CH_2CH_2CH_2CH_2CH_3$	$CH_3CH_2CH_2CH_2CH_2OH$	$HOCH_2CH_2CH_2CH_2OH$
Hexane	1-Pentanol	1,4-Butanediol
(insoluble)	(2.3 g/mL water)	(infinitely soluble)

14.63 Each can be prepared from 2-methyl-1-propanol (circled) as shown in this flow chart.

$$\underset{\underset{OH}{|}}{\overset{CH_3}{\overset{|}{CH_3CCH_3}}} \xleftarrow[H_2SO_4]{H_2O} \overset{CH_3}{\overset{|}{CH_3C{=}CH_2}} \xleftarrow[-H_2O]{H_2SO_4} \boxed{\overset{CH_3}{\overset{|}{CH_3CHCH_2OH}}} \xrightarrow[H_2SO_4]{K_2Cr_2O_7} \overset{CH_3}{\overset{|}{CH_3CHCOOH}}$$

(b) (a) (c)

14.65 (a) The two functional groups are a carboxyl group and a disulfide group.
(b) Its reduction product is dihydrolipoic acid.

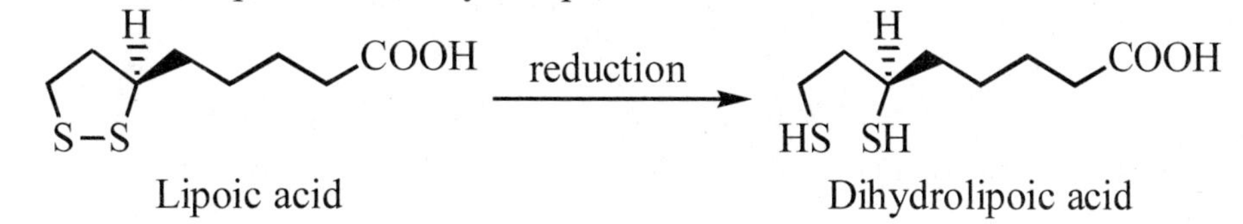

Lipoic acid Dihydrolipoic acid

15.1 The enantiomers of each part are drawn with two groups in the plane of the paper, a third group toward you in front of the plane, and the fourth group away from you behind the plane.

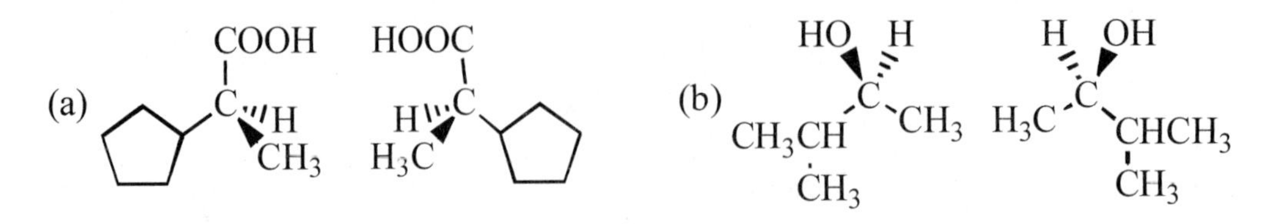

15.3 The order of priorities and the configuration is shown in the drawings.

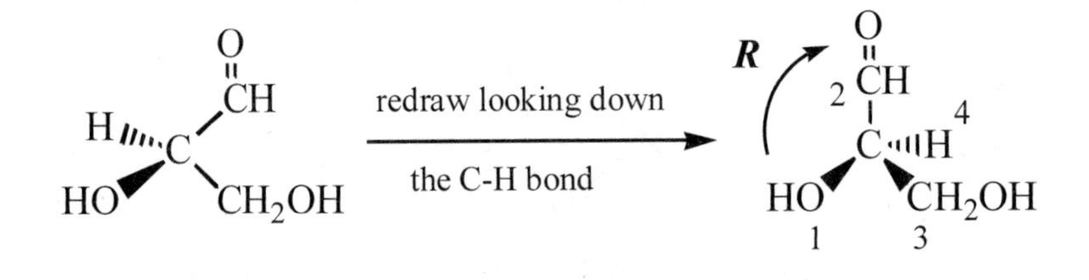

15.5 Four stereoisomers are possible for 3-methylcyclohexanol. The *cis* isomers are one pair of enantiomers and the *trans* isomers are a second pair of enantiomers.

15.7 Chirality is a property of an object. The object is not superposable on its mirror image; that is, the object has handedness. 2-Butanol is a chiral molecule.

15.9 Stereoisomers are isomers that have the same molecular formula and the same connectivity, but a different orientation of their atoms in space. Three examples are *cis-trans* isomers, enantiomers, and diastereomers.

15.11 (a) Chiral. (b) Achiral. (c) Achiral. (d) Achiral.
(e) Chiral, unless you are on the equator, in which case it goes straight down and has no chirality.

15.13 Neither *cis*-2-butene nor *trans*-2-butene is chiral. Each is superposable on its mirror image.

15.15 Compounds (a), (c), and (d) contain stereocenters, here identified with asterisks.

(a) $CH_3\overset{Cl}{\underset{*}{C}}HCH_2CH_2CH_3$ (c) $CH_2{=}CH\overset{Cl}{\underset{*}{C}}HCH_3$ (d) $\overset{Cl}{C}H_2\overset{Cl}{\underset{*}{C}}HCH_3$

15.17 The stereocenter in each chiral compound is identified with an asterisk.

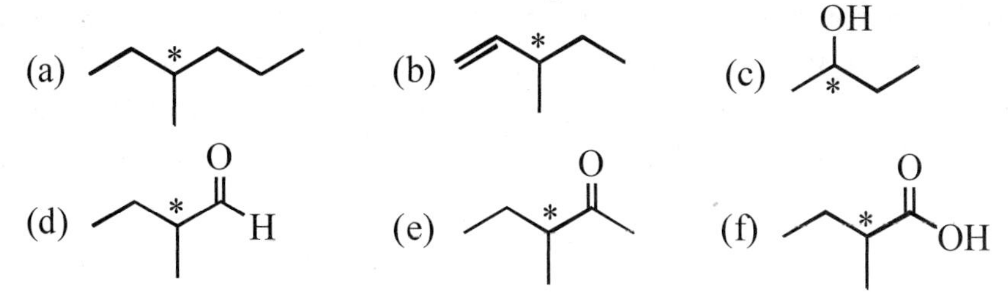

15.19 The following pairs of structures are nonsuperposable mirror images.

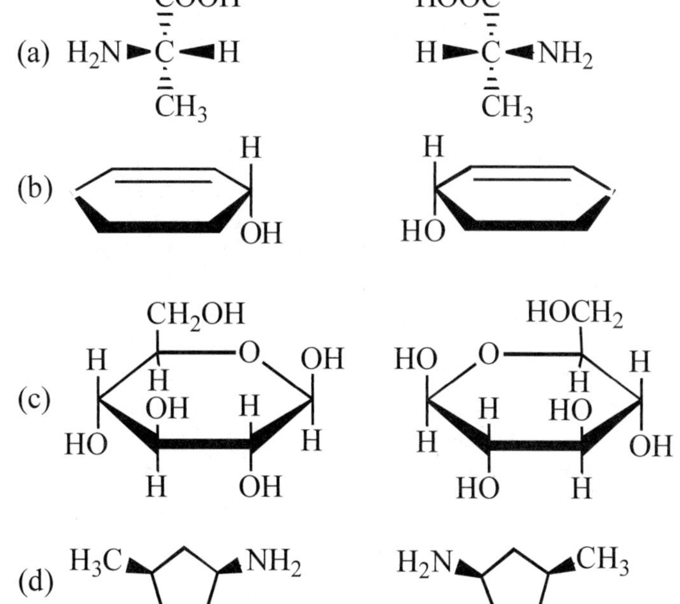

15.21 Parts (b) and (c) contain stereocenters.

(b) HO OH * (c) OH *

15.23 Each stereocenter is identified with an asterisk. Under each structural formula is the number of stereoisomers possible. Compound (b) has no stereocenter.

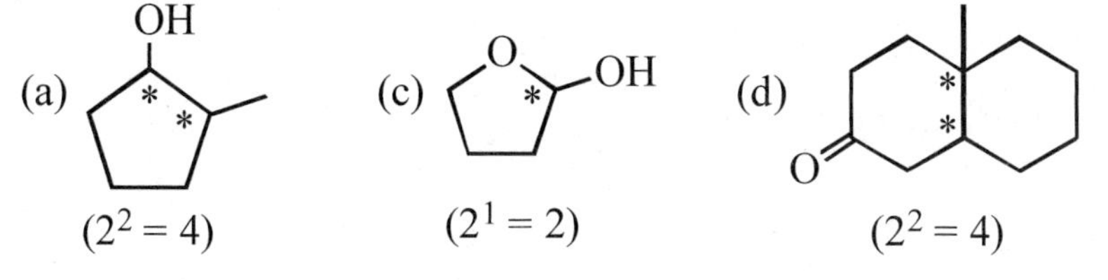

($2^2 = 4$) ($2^1 = 2$) ($2^2 = 4$)

15.25 The optical rotation of its enantiomer is +41°.

15.27 To say that a drug is chiral means that it has handedness - that it has one or more stereocenters and the possibility for two or more stereoisomers. Just because a compound is chiral does not mean that it will be optically active. I t may be chiral and be present as a racemic mixture, in which case it will have no effect on the plane of polarized light. If, however, it is present as a single enantiomer, it will rotate the plane of polarized light.

15.29 The two stereocenters are identified with asterisks.

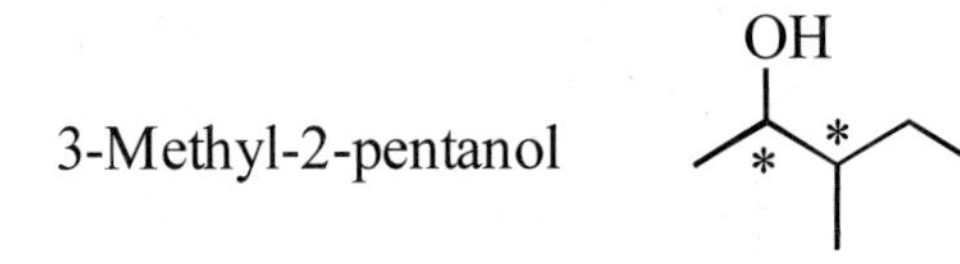

15.31 Each stereocenter is identified with an asterisk. Under the name of each compound is the number of stereoisomers possible for it.

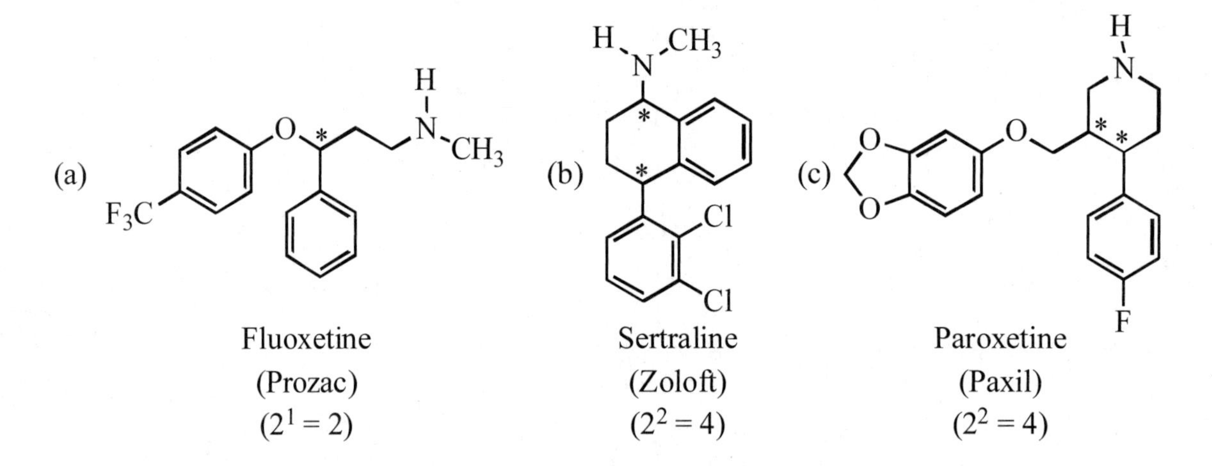

15.33 (a) The only possible compound that does not show *cis-trans* isomerism and has no stereocenter is 1-methylcyclohexanol.

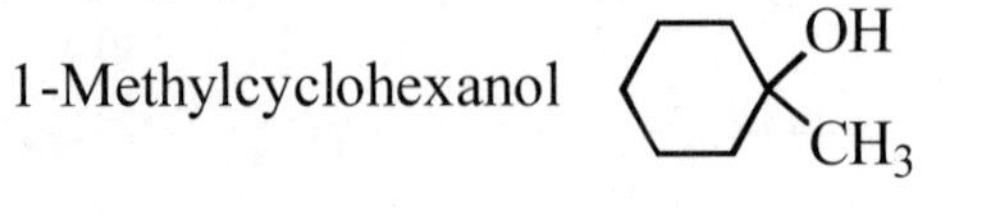

(b) Only 4-methylcyclohexanol shows *cis-trans* isomerism but has no stereocenter. Drawn here are the *cis* and *trans* isomers.

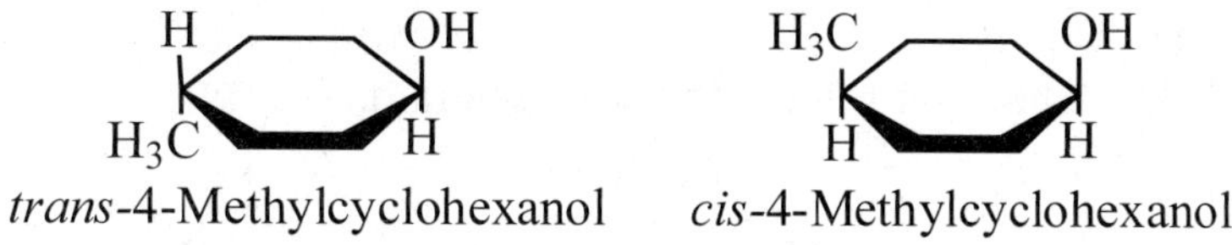

trans-4-Methylcyclohexanol *cis*-4-Methylcyclohexanol

(c) Both 2-methylcyclohexanol and 3-methylcyclohexanol have two stereocenters and can exist as *cis* and *trans* isomers. The stereocenters in each are identified with asterisks.

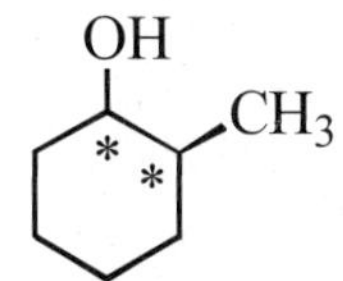

2-Methylcyclohexanol 3-Methylcyclohexanol

15.35 The majority of spirals have a right-handed twist because the machines that make them all impart the same twist.

15.37 (a) In this chair conformation of glucose, carbons 1, 2, 3, 4, and 5 are stereocenters.
(b) There are $2^5 = 32$ stereoisomers possible.
(c) Because enantiomers always occur in pairs, there are 16 pairs of enantiomers possible.

16.1 Pyrrolidine has nine hydrogens; its molecular formula is C_4H_9N. Purine has four hydrogens; its molecular formula is $C_5H_4N_4$.

16.3 Following is a line-angle formula for each compound.

(a) HO$\sim$$NH_2$ (b) Ph–NH–Ph (c) (isopropyl)–NH–(isopropyl)

16.5 The product of each reaction is an amine salt.

(a) $(CH_3CH_2)_3\overset{+}{N}H\ Cl^-$ (b) piperidinium ($\overset{+}{N}H_2$) $CH_3\overset{O}{\overset{\|}{C}}O^-$

16.7 In an aliphatic amine, all carbon groups bonded to nitrogen are alkyl groups. In an aromatic amine, one or more of the carbon groups bonded to nitrogen are aryl (aromatic) groups.

16.9 Following is a structural formula for each amine:

(a) NH_2 (b) $CH_3(CH_2)_6CH_2NH_2$ (c) NH_2

(d) $H_2N(CH_2)_5NH_2$ (e) NH_2, Br (f) $(CH_3CH_2CH_2CH_2)_3N$

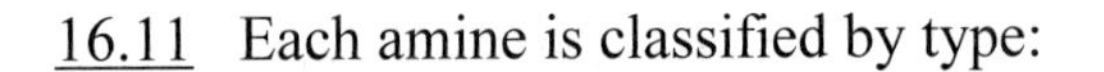

16.11 Each amine is classified by type:

(a)

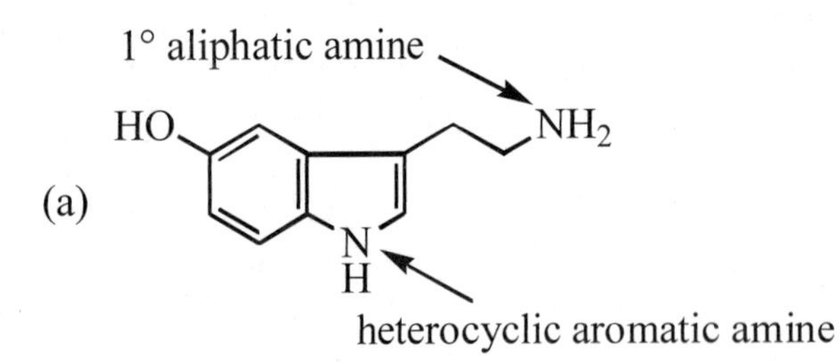

(b)

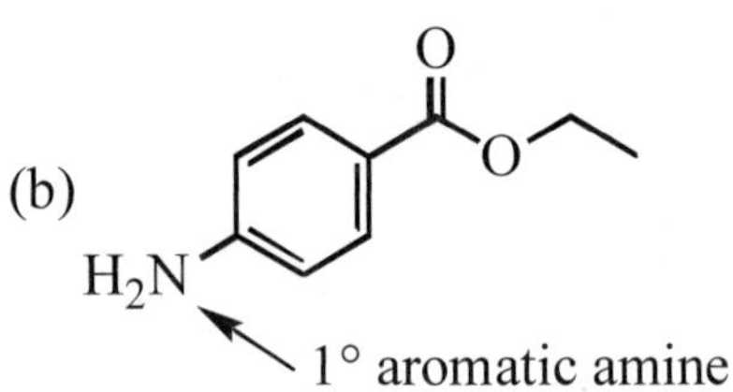

(c)

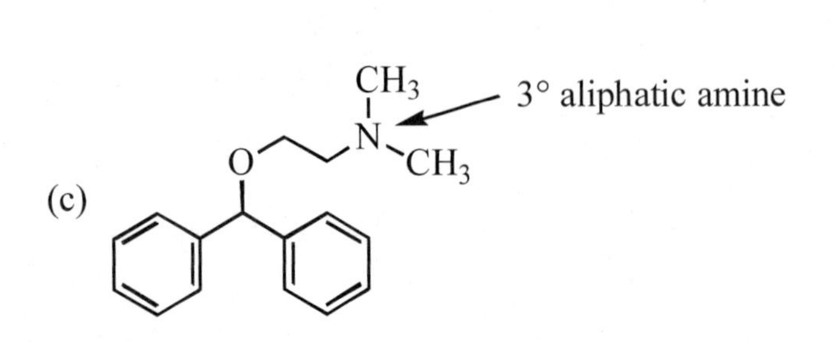

16.13 There are four primary amines of this molecular formula, three secondary amines, and one tertiary amine. Only 2-butanamine is chiral.

1° amines:

NH_2

1-Butanamine
(Butylamine)

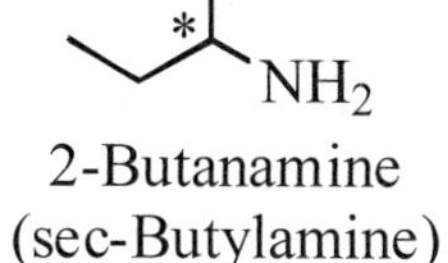

2-Butanamine
(sec-Butylamine)

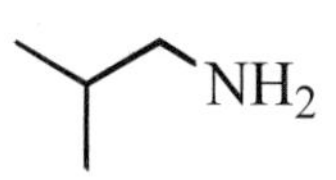

2-Methyl-1-
propanamine
(Isobutylamine)

NH_2

1,1-Dimethyl-
ethanamine
(*tert*-Butylamine)

2° amines:

H N CH_3

Methylpropyl-
amine

H N CH_3

Methylisopropyl-
amine

H N

Diethylamine

3° amine:

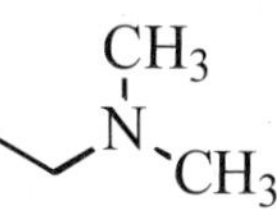

Dimethylethylamine

16.15 Both propylamine (a 1° amine) and ethylmethylamine (a 2° amine) have an N-H group and hydrogen bonding occurs between their molecules in the liquid state. Because of this intermolecular force of attraction, these two amines have higher boiling points than trimethylamine, which has no N-H bond and, therefore, cannot participate in intermolecular hydrogen bonding.

16.17 2-Methylpropane is a nonpolar hydrocarbon and the only attractive forces between its molecules in the liquid state are the very weak London dispersion forces. Both 2-propanol and 2-propanamine are polar molecules and associate in the liquid state by hydrogen bonding. Hydrogen bonding is stronger between alcohol molecules than between amine molecules because of the greater strength of an O-H----O hydrogen bond compared to an N-H----N hydrogen bond. It takes more energy (a higher temperature) to separate an alcohol molecule in the liquid state from its neighbors than to separate an amine molecule from its neighbors and, therefore, the alcohol has the higher boiling point.

16.19 Nitrogen is less electronegative than oxygen and, therefore, more willing to donate its unshared pair of electrons to H^+ in an acid-base reaction to form a salt, thus making the amine a stronger base.

16.21 (a) Ethylammonium chloride (b) Diethylammonium chloride
(c) Anilinium hydrogen sulfate

16.23 The form of amphetamine present at both pH 1.0 and pH 7.4 is the ammonium ion shown in answer to Problem 16.25(b).

16.25 Following are completed equations.

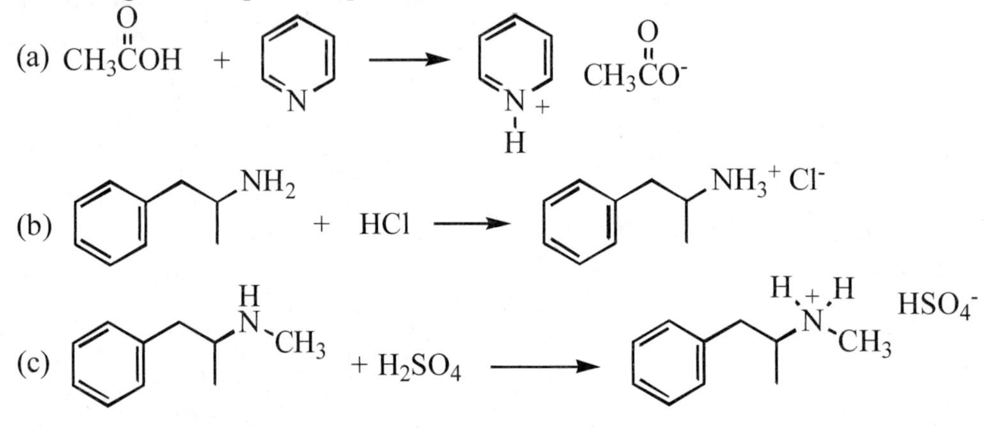

16.27 (a) Tamoxifen contains three aromatic (benzene) rings, one carbon-carbon double bond, one ether, and one amine.
(b) The amine is a 3° amine.
(c) Two stereoisomers are possible, a pair of *cis-trans* isomers.

16.29 (a) Epinephrine has two phenolic –OH groups, whereas amphetamine has none. Epinephrine has a 2° alcohol on its carbon side chain whereas amphetamine has none. Epinephrine is a 2° amine, whereas amphetamine is a 1° amine. Finally, both epinephrine and amphetamine are chiral, but their stereocenters are in different locations within the molecule.
(b) Methamphetamine is a 2° aliphatic amine whereas amphetamine is a 1° aliphatic amine. Both compounds are chiral at the same carbon.

16.31 Alkaloids are basic nitrogen-containing compounds found in the roots, bark, leaves, berries, or fruits of plants. In almost all alkaloids, the nitrogen atom is present as a member of a ring; that is, it is present in a heterocyclic ring. By definition, alkaloids are nitrogen-containing bases and, therefore, turn red litmus blue.

16.33 The tertiary aliphatic amine in the five-membered ring is the stronger base.

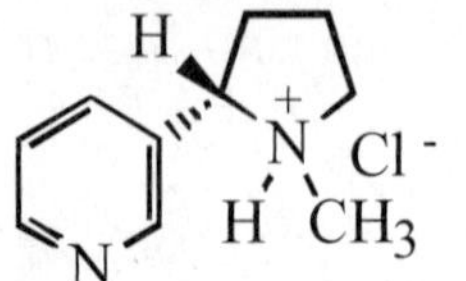

16.35 The common structural feature is a benzene ring fused to a seven-membered ring containing two nitrogen atoms. This parent structural feature is named benzodiazepine.

16.37 The structural formula of 4-aminobutanoic acid is drawn on the left showing the 1° amino group (a base) and the carboxyl group (an acid). It is drawn on the right as an internal salt.

a primary amine → $H_2NCH_2CH_2CH_2C(=O)OH$ ← a carboxylic acid

un-ionized amino and carboxyl groups

$H_3\overset{+}{N}CH_2CH_2CH_2C(=O)O^-$

internal salt

16.39 Their salts are more soluble in water and in body fluids, and are more stable (less reactive) toward oxidation by atmospheric oxygen.

16.41 In order of decreasing ability to form intermolecular hydrogen bonds, they are CH_3OH > $(CH_3)_2NH$ > CH_3SH. An O-H bond is more polar than an N-H bond, which is more polar than an S-H bond.

16.43 Butane, the least polar molecule, has the lowest boiling point and 1-propanol, the most polar molecule, has the highest boiling point.

$CH_3CH_2CH_2CH_3$	$CH_3CH_2CH_2NH_2$	$CH_3CH_2CH_2OH$
-0.5°C	48°C	97°C

16.45 The alcohol will be unchanged. The amine will react with HCl to form a salt.

$$CH_3CH_2CH_2NH_3^+\ Cl^-$$

16.47 (a) Procaine is achiral; it has no stereocenter.
(b) The 3° aliphatic amine is the stronger base.
(c) Following is a structural formula for its hydrochloride salt.

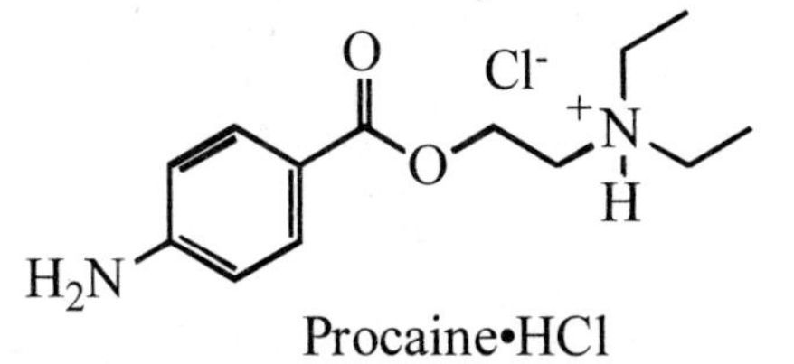

Procaine•HCl

16.49 (a) The amino group is a 3° aliphatic amine.
(b) The three stereocenters are identified with asterisks.

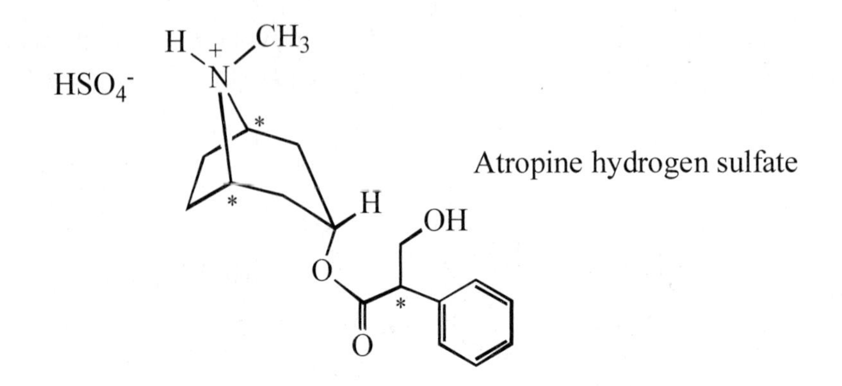

(c) Because it is an ammonium salt, atropine sulfate is more soluble in water than atropine.
(d) Dilute aqueous solutions of atropine are basic because its 3° aliphatic amine reacts with water to produce hydroxide ions.

16.51 Structural formula A contains both an acid (the carboxyl group) and a base (the 1° amino group). The acid-base reaction between them gives structural formula B, which is the better representation of this amino acid.

17.1 (a) 3,3-Dimethylbutanal (b) Cyclopentanone (c) 1-Phenyl-1-propanone

17.3 (a) 2,3-Dihydroxypropanal (b) 2-Aminobenzaldehyde (c) 5-Amino-2-pentanone

17.5 Each primary alcohol comes from reduction of an aldehyde. Each secondary alcohol comes from reduction of a ketone.

(a) =O (b) CH_3O–⟨⟩–$CH_2\overset{O}{\overset{\|}{C}}H$ (c) O O

17.7 (a) A hemiacetal formed from 3-pentanone (a ketone) and ethanol.
(b) Neither a hemiacetal nor an acetal. This compound is the dimethyl ether of ethylene glycol.
(c) An acetal derived from 5-hydroxypentanal and methanol.

17.9 The carbonyl carbon of an aldehyde is bonded to at least one hydrogen. The carbonyl carbon of a ketone is bonded to two carbon groups.

17.11 No. To be a carbon stereocenter, the carbon atom must have four different groups bonded to it. The carbon atom of a carbonyl group has only three groups bonded to it.

17.13 (a) Cortisone contains three ketones, one 3° alcohol, one 1° alcohol, and one carbon-carbon double bond. Aldosterone contains two ketones, one aldehyde, one 1° alcohol, one 2° alcohol, and one carbon-carbon double bond.
(b) The stereocenters are identified with asterisks.

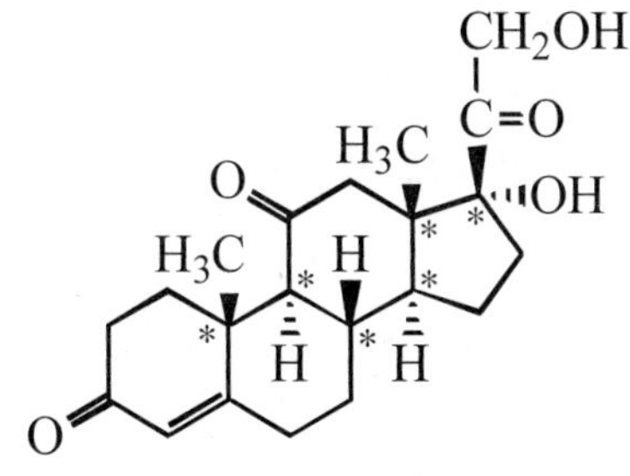

Cortisone

$2^6 = 64$ possible stereoisomers

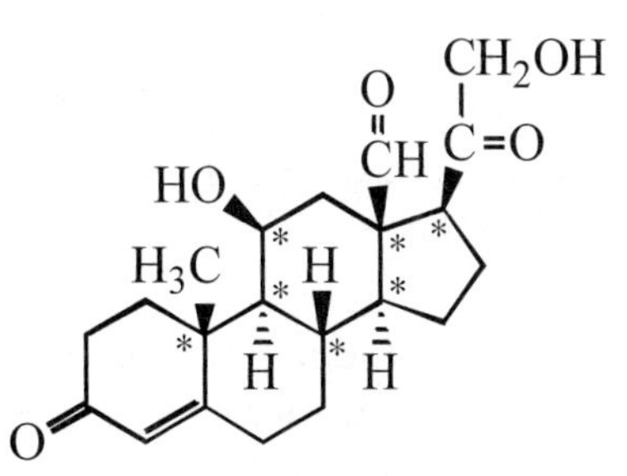

Aldosterone

$2^7 = 128$ possible stereoisomers

17.15 Following are structural formulas for each aldehyde.

(a) 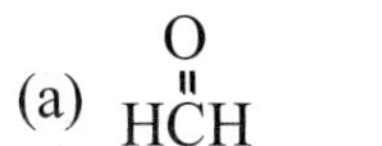

(b) O H

(c) 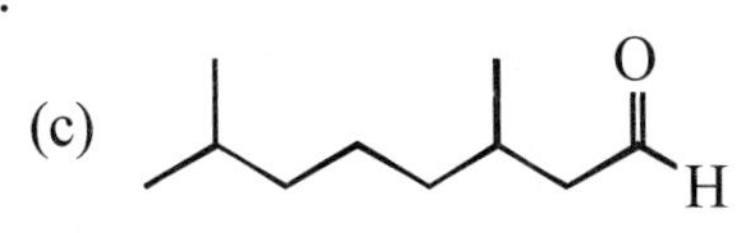

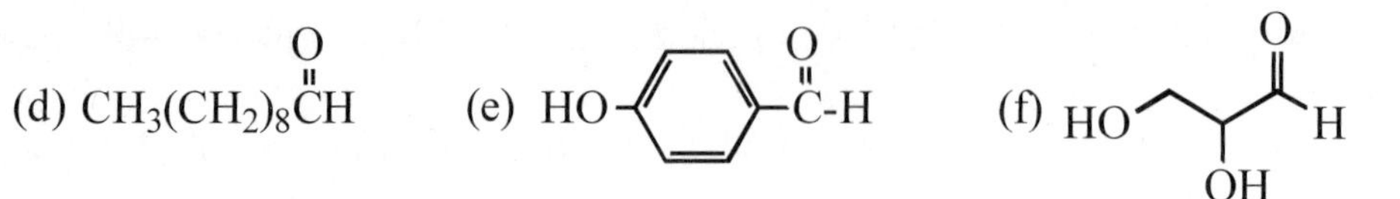

17.17 (a) 4-heptanone
(b) 2-Methylcyclopentanone
(c) *cis*-2-Methyl-2-butenal
(d) 2-Hydroxypropanal
(e) 1-Phenyl-2-propanone
(f) Hexanedial

17.19 (a) The chain is not numbered correctly. Its name is 2-butanone.
(b) The compound is an aldehyde. Its name is butanal.
(c) The longest chain is five carbons. Its name is pentanal.
(d) The location of the ketone takes precedence. Its name is 3,3-dimethyl-2-butanone.

17.21 When comparing the boiling points of different compounds, boiling points increase as the intermolecular forces increase. This trend occurs because more energy is required to break the stronger intermolecular forces than the energy required to break weaker intermolecular forces.
(a) Ethanol has the higher boiling point because its hydrogen bonding intermolecular forces are stronger than the weaker dipole-dipole intermolecular forces present in acetaldehyde.
(b) 3-Pentanone has the higher boiling point because the London dispersion forces in the larger ketone, 3-pentanone, are greater than those present in the smaller ketone, acetone.
(c) Butanal has the higher boiling point because its dipole-dipole intermolecular forces are stronger than the London dispersion forces present in butane.
(d) 2-Butanol has the higher boiling point because its hydrogen bonding intermolecular forces are stronger than the weaker dipole-dipole intermolecular forces present in butanone.

17.23 Acetone has the higher boiling point than ethyl methyl ether because the intermolecular dipole-dipole attractive forces between the carbonyl groups of acetone molecules is greater than the dipole-dipole attraction between ethyl methyl ether molecules.

17.25 Acetaldehyde is a hydrogen bond acceptor and forms hydrogen bonds with water primarily through its carbonyl oxygen.

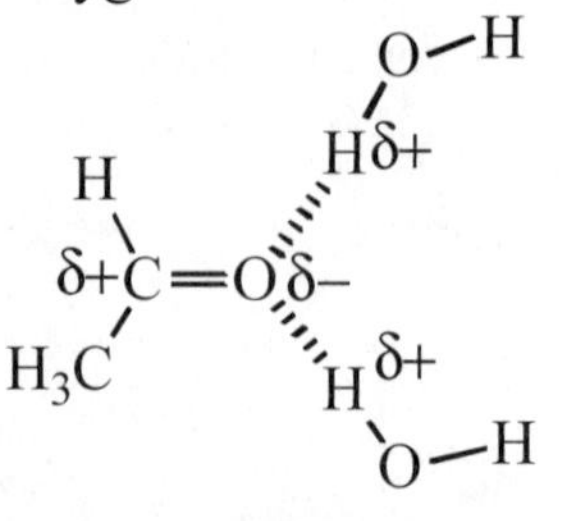

17.27 Aldehydes are oxidized by potassium dichromate in sulfuric acid to carboxylic acids. Ketones are not oxidized under these conditions. Secondary alcohols are oxidized to ketones.

(a) $CH_3CH_2CH_2\overset{O}{\overset{\|}{C}}HOH$ (b) $C_6H_5\overset{O}{\overset{\|}{C}}OH$ (c) no reaction (d) cyclohexanone (ring=O)

17.29 (a) Treat each with Tollens' reagent. Only pentanal will give a silver mirror.
(b) Treat each with $K_2Cr_2O_7/H_2SO_4$. Only 2-pentanol is oxidized (to 2-pentanone), which causes the red color of $Cr_2O_7^{2-}$ ion to disappear and be replaced by the green color of Cr^{3+} ion.

17.31 The white solid is benzoic acid, formed by air oxidation of benzaldehyde.

17.33 These experimental conditions reduce an aldehyde to a primary alcohol and a ketone to a secondary alcohol. Both (a) and (c) are chiral.

(a) $CH_3\overset{OH}{\overset{|}{C}}HCH_2CH_3$ (b) $CH_3(CH_2)_4CH_2OH$ (c) cyclopentane ring bearing OH and CH_3 (d) benzene ring bearing CH_2OH and OH

17.35 (a) The following is the structural formula for 1,3-dihydroxy-2-propanone.

$$HOCH_2\overset{O}{\overset{\|}{C}}CH_2OH$$

1,3-Dihydroxy-2-propanone
(Dihydroxyacetone)

(b) Because it has two hydroxyl groups and one carbonyl group, all of which can interact with water molecules by hydrogen bonding, predict that it is soluble in water.
(c) Its reduction gives 1,2,3-propanetriol, better known as glycerol or glycerin.

$$HOCH_2\overset{O}{\overset{\|}{C}}CH_2OH \xrightarrow[H_2O]{NaBH_4} HOCH_2\overset{OH}{\overset{|}{C}}HCH_2OH$$

1,2,3-Propanetriol

17.37 The first two reactions are reduction. There is no reaction in (c) or (d); ketones are not oxidized by these reagents.

(a,b) 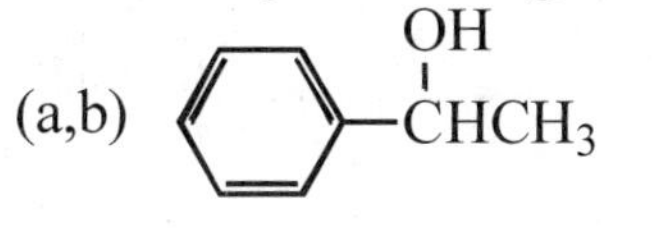

17.39 Compound (a) has one enol. Compounds (b) and (c) have two enols each.

(a) $CH_3CH{=}\overset{OH}{C}H$ (b) $CH_3\overset{OH}{C}{=}CHCH_3$ and $CH_2{=}\overset{OH}{C}CH_2CH_3$

(c) OH CH_3 and OH CH_3

17.41 A hemiacetal contains a carbon atom bonded to one –OH group and one –OR group, where R may be an alkyl or aryl group. An acetal contains a carbon atom bonded to two –OR groups, where R may be alkyl or aryl.

17.43 (a) Hemiacetal (b) Acetal (c) Neither
(d) Hemiacetal (e) Acetal (f) Acetal

17.45 Following are structural formula for the product of each hydrolysis.

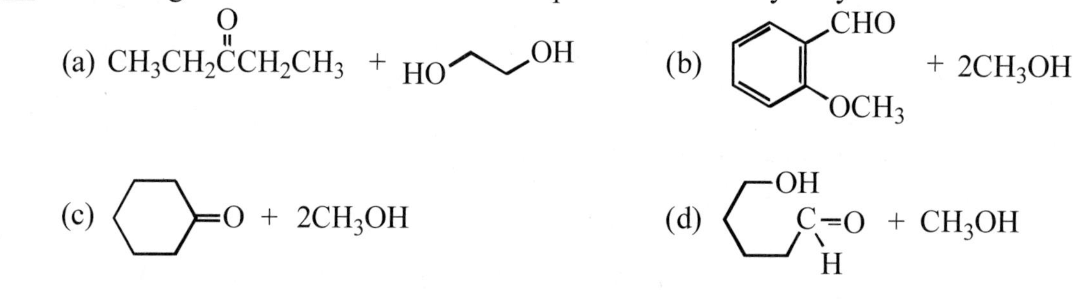

17.47 *Hydration* refers to the addition of one or more molecules of water to a substance. An example of hydration is the acid-catalyzed addition of water to propene to give 2-propanol. *Hydrolysis* refers to the reaction of a substance with water, generally with the breaking (lysis) of one or more bonds in the substance. An example of hydrolysis is the acid-catalyzed reaction of an acetal with a molecule of water to give an aldehyde or ketone and two molecules of alcohol.

17.49 The flow chart shows the synthesis of each target molecule.

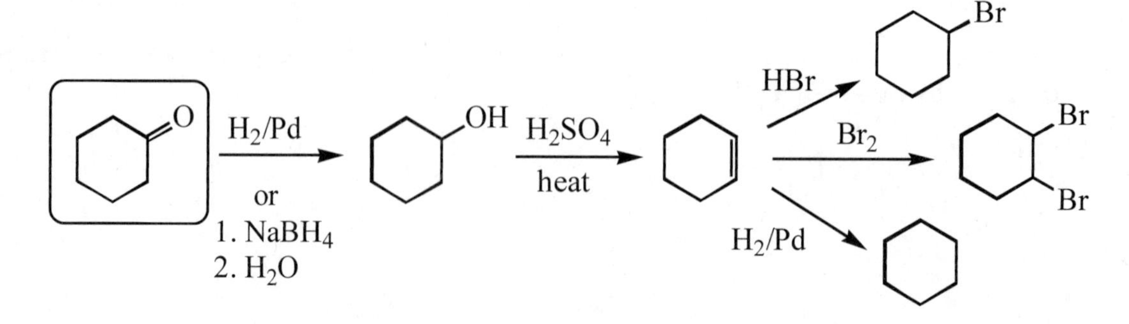

17.51 Compounds (a), (b), and (d) can be formed by reduction of the aldehyde or ketone shown. Compound (c) is a 3° alcohol and cannot be formed in this manner.

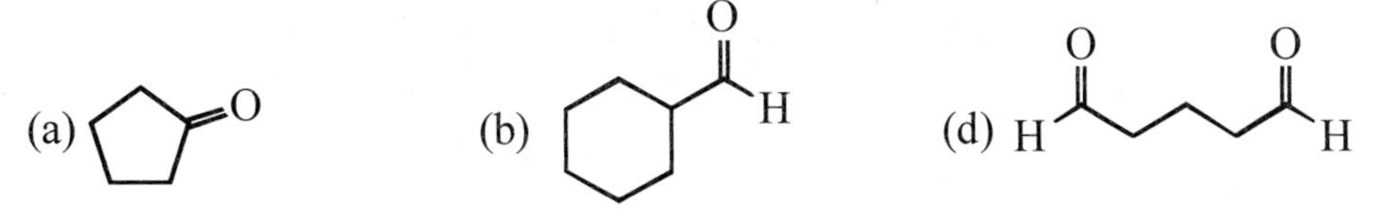

17.53 Reagents for each conversion are shown over the reaction arrow.

(a) $C_6H_5C(=O)CH_2CH_3 \xrightarrow[\text{or } 1.\ NaBH_4;\ 2.\ H_2O]{H_2/Pd} C_6H_5CH(OH)CH_2CH_3 \xrightarrow[\text{heat}]{H_2SO_4} C_6H_5CH{=}CHCH_3$

(b) cyclopentanone $\xrightarrow[\text{or } 1.\ NaBH_4;\ 2.\ H_2O]{H_2/Pd}$ cyclopentanol (–OH) $\xrightarrow[\text{heat}]{H_2SO_4}$ cyclopentene $\xrightarrow{HCl}$ chlorocyclopentane (–Cl)

17.55 (a) Each compound is insoluble in water. Treat each with dilute aqueous HCl. Aniline, an aromatic amine, reacts with HCl to form a water-soluble salt. Cyclohexanone does not react with this reagent and is insoluble in aqueous HCl.

(b) Treat each with a solution of Br_2/CH_2Cl_2. Cyclohexene reacts to discharge the red color of Br_2 and to form 1,2-dibromocyclohexane, a colorless compound. Cyclohexanol does not react with this reagent.

(c) Treat each with a solution of Br_2/CH_2Cl_2. Cinnamaldehyde, which contains a carbon-carbon double bond, reacts to discharge the red color of Br_2 and to form 2,3-dibromo-3-phenylpropanal, a colorless compound. Benzaldehyde does not react with this reagent.

17.57 Using this reducing agent, each aldehyde is converted to a 1° alcohol and each ketone is converted to a 2° alcohol.

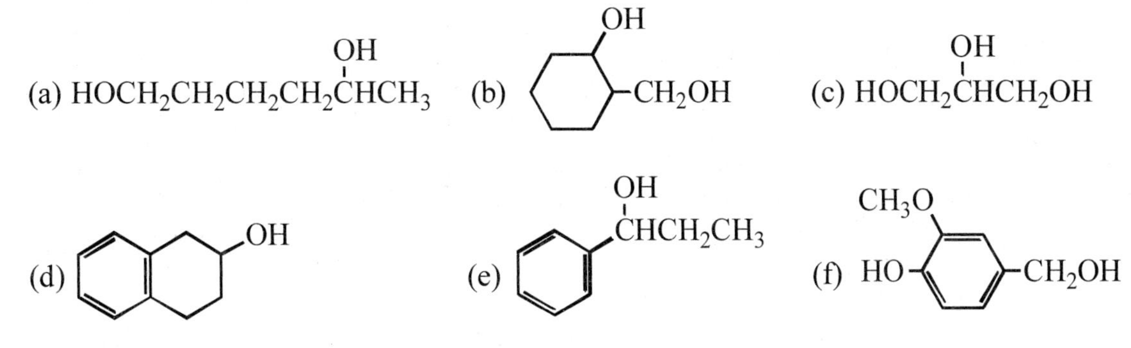

17.59 Following is a structural formula for each compound.

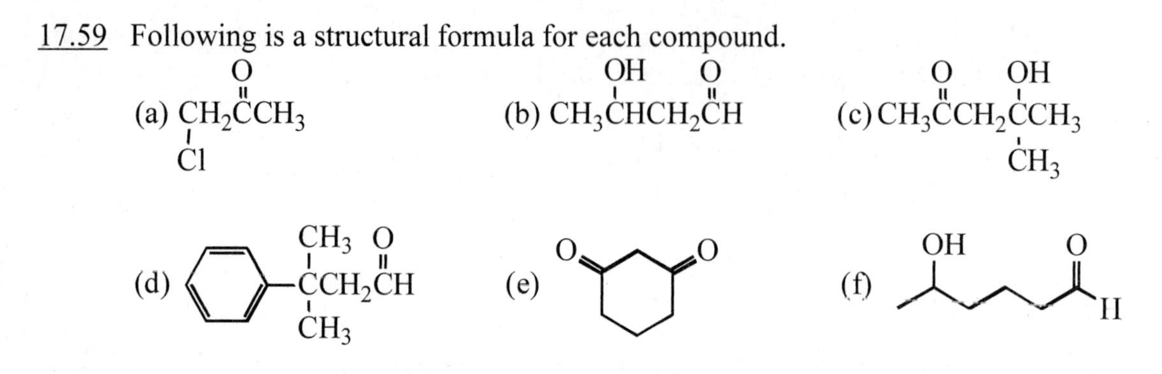

17.61 The intermolecular attraction due to hydrogen bonding between molecules of 1-propanol is stronger than the attraction between molecules of propanal.

17.63 (a) The hydroxyaldehyde is first redrawn to show the OH group nearer the CHO group. Closing the ring in hemiacetal formation gives the cyclic hemiacetal.
(b) 5-Hydroxyhexanal has one stereocenter, and two stereoisomers (one pair of enantiomers) are possible.
(c) The cyclic hemiacetal has two stereocenters, and four stereoisomers (two pairs of enantiomers) are possible.

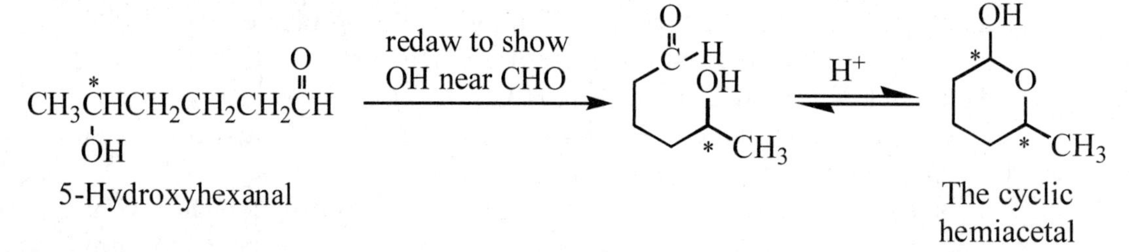

17.65 Alkenes undergo acid-catalyzed hydration in aqueous H_2SO_4 to give alcohols. Aldehydes and ketones give alcohols upon reduction using either $NaBH_4$ or H_2 in the presence of a transition metal catalyst. The following are the alkene and aldehyde/ketone starting materials that yield the desired alcohols under the appropriate conditions.

(a) $CH_2{=}CH_2$ $\quad CH_3\overset{O}{\overset{\|}{C}}H$

(b)

(c) $CH_2{=}CHCH_3$ $\quad CH_3\overset{O}{\overset{\|}{C}}CH_3$

(d) $-CH{=}CH_2$ $\quad -\overset{O}{\overset{\|}{C}}CH_3$

17.67 (a) Carbon 4 provides the –OH group and carbon 1 provides the –CHO group.
(b) Following is a structural formula for the free aldehyde.

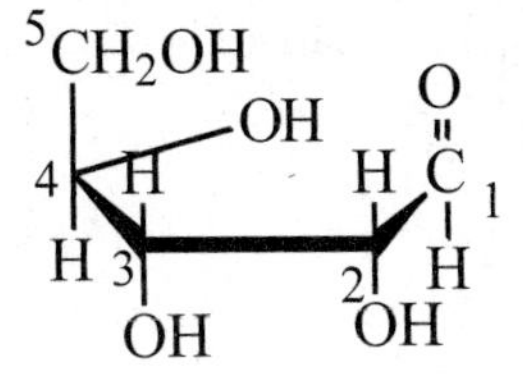

17.69 Each compound reduces a carbonyl group of an aldehyde or ketone to an alcohol by delivering a hydride ion (H:⁻) to the carbonyl carbon. The hydrogen of the resulting –OH group comes from an –OH group of the solvent,

17.71

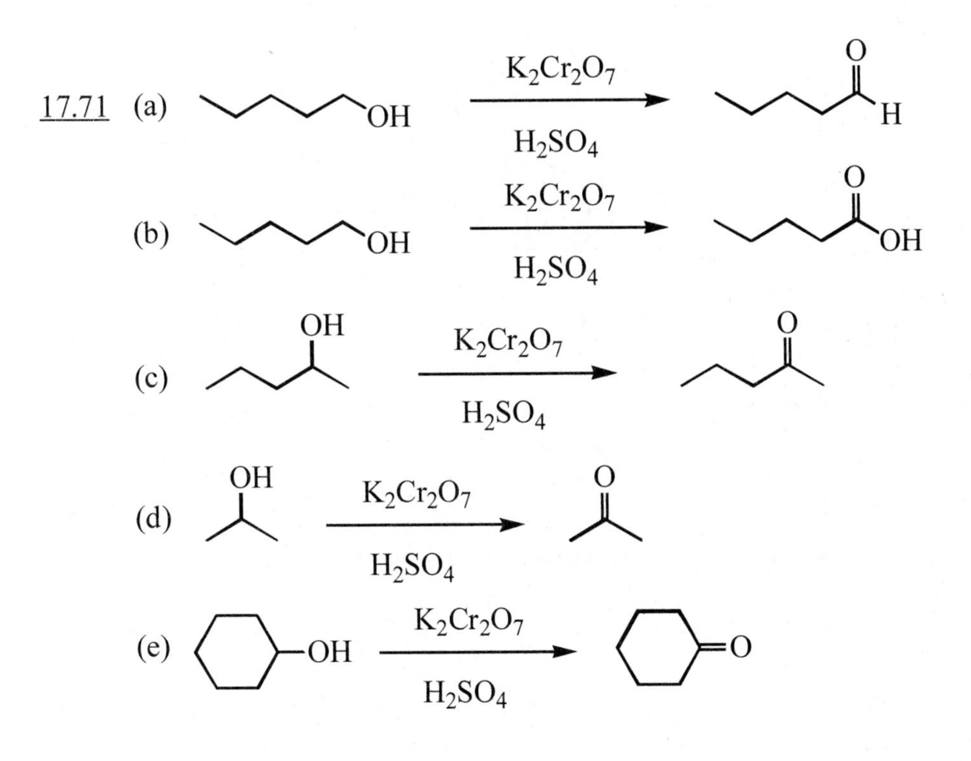

18.1 (a) 2,3-Dihydroxypropanoic acid (b) 3-Aminopropanoic acid
(c) 3,5-Dihydroxy-3-methylpentanoic acid
Glyceric acid and mevalonic acid are chiral.

18.3 (a) O O + H_2O (b) O O + H_2O

18.5 (a) 3,4-Dimethylpentanoic acid (b) 2-Aminobutanoic acid (c) Hexanoic acid

18.7 Following are structural formulas for each carboxylic acid.

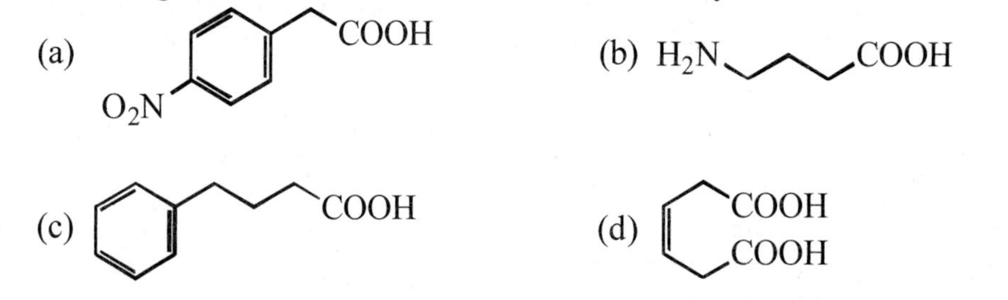

18.9 Following are structural formulas for each salt.

(a) $C_6H_5\overset{O}{\overset{\|}{C}}O^- Na^+$ (b) $CH_3\overset{O}{\overset{\|}{C}}O^- Li^+$ (c) $CH_3\overset{O}{\overset{\|}{C}}O^- NH_4^+$

(d) $Na^+\ ^-O\overset{O}{\overset{\|}{C}}(CH_2)_4\overset{O}{\overset{\|}{C}}O^- Na^+$ (e) $\overset{O}{\overset{\|}{C}}O^- Na^+$, OH (f) $(CH_3CH_2CH_2\overset{O}{\overset{\|}{C}}O^-)_2Ca^{2+}$

18.11 One of the carboxyl groups in this salt is present as –COO⁻, the other as –COOH.

$$HO\overset{O}{\overset{\|}{C}}-\overset{O}{\overset{\|}{C}}O^- K^+$$

18.13 If you draw this molecule correctly to show this internal hydrogen bonding, you will see that the hydrogen-bonded part of the molecule forms a six-membered ring.

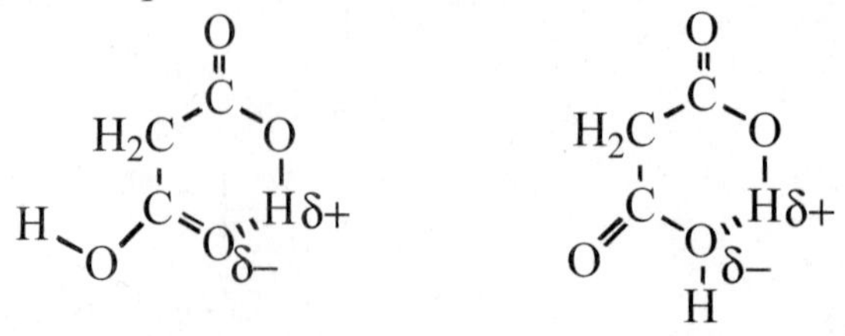

18.15 Propanoic acid has the higher boiling point (141°C). It can participate in intermolecular hydrogen bonding through its carboxyl group. This intermolecular attractive force must be overcome before a propanoic acid molecule can escape from the liquid phase to the vapor phase. There is no comparable intermolecular attractive force between molecules of methyl acetate in the liquid state.

18.17 In order of increasing boiling point, they are diethyl ether (b.p. 35°C), 1-butanol (b.p. 117°C), and propanoic acid (b.p. 141°C). Boiling points increase as the degree of intermolecular hydrogen bonding increases.

18.19 Carboxylic acids are the stronger acids, with pK_a values in the range 4.0 - 5.0. Phenols are the weaker acids, with pK_a values of approximately 10.0.

18.21 Following are equations for these acid-base reactions.

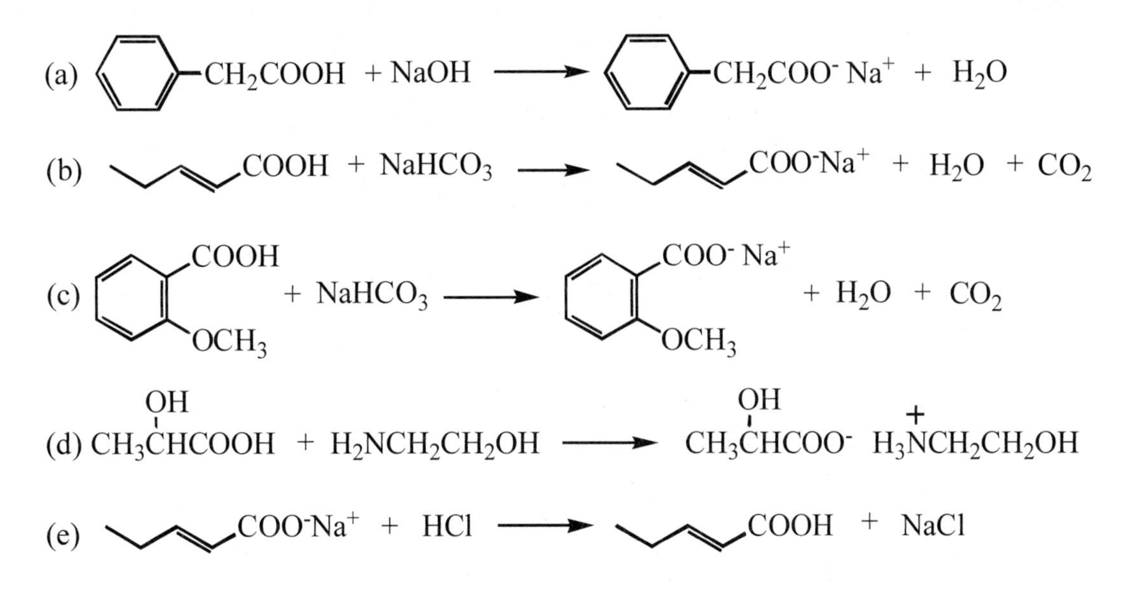

18.23 The acid-base reaction is the neutralization of formic acid by sodium bicarbonate.

$$\underset{\text{Formic acid}}{HCOOH} + NaHCO_3 \longrightarrow \underset{\text{Sodium formate}}{HCOO^-Na^+} + H_2O + CO_2$$

18.25 Using $\frac{K_a}{[H_3O^+]} = \frac{[A^-]}{[HA]}$ and substituting the following values: $\frac{[A^-]}{[HA]} = \frac{10^{-5}}{[10^{-pH}]}$;
the following table can be generated:

pH	$\frac{[A^-]}{[HA]}$
(a) 2.0	10^3
(b) 5.0	1.0
(c) 7.0	10^{-2}
(d) 9.0	10^{-4}
(e) 11.0	10^{-6}

18.27 The pK_a of lactic acid is 4.07. At this pH, lactic acid would be present as 50% $CH_3CH(OH)COOH$) and 50% $CH_3CH(OH)COO^-$. At pH 7.35 to 7.45, which is more basic than pH 4.07, lactic acid would be present as $CH_3CH(OH)COO^-$.

18.29 In part (a), the –COOH group is a stronger acid than the $-NH_3^+$ group.

(a) $CH_3CH(NH_3^+)COOH + NaOH \longrightarrow CH_3CH(NH_3^+)COO^-Na^+ + H_2O$

(b) $CH_3CH(NH_3^+)COO^-Na^+ + NaOH \longrightarrow CH_3CH(NH_2)COO^-Na^+ + H_2O$

18.31 In part (a), the amine is the stronger base.

(a) $CH_3CH(NH_2)COO^-Na^+ + HCl \longrightarrow CH_3CH(NH_3^+)COO^-Na^+ + Cl^-$

(b) $CH_3CH(NH_3^+)COO^-Na^+ + HCl \longrightarrow CH_3CH(NH_3^+)COOH + NaCl$

18.33 Following is a structural formula for the ester formed in each reaction. Water is also formed along with each indicated ester.

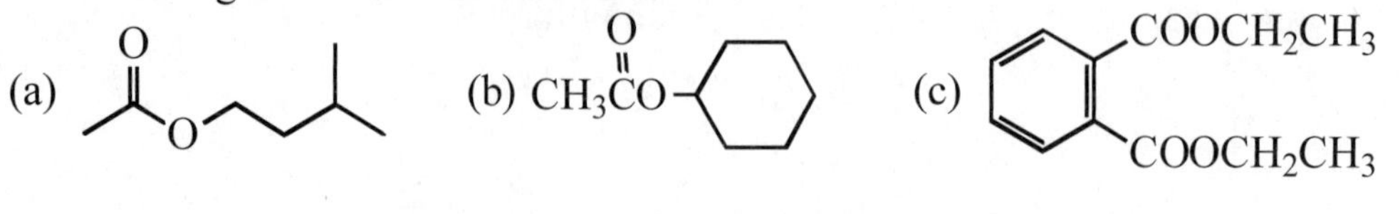

18.35 Following is a structural formula for methyl 2-hydroxybenzoate.

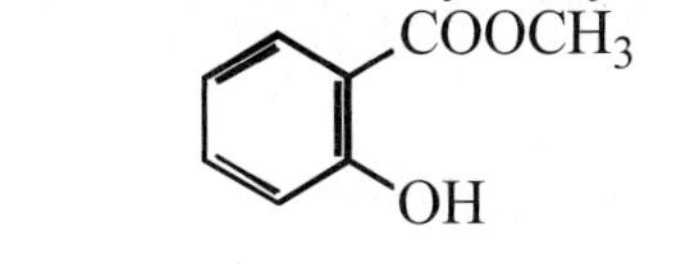

18.37 Both (e) and (g) give no reaction. Phenylacetic acid is unreactive toward $NaBH_4$ in (e) and Ni/H_2 in (g).

(a) $C_6H_5CH_2COO^-Na^+ + CO_2 + H_2O$

(b) $C_6H_5CH_2COO^-Na^+ + H_2O$

(c) $C_6H_5CH_2COO^-NH_4^+$

(d) $C_6H_5CH_2CH_2OH$

(f) $C_6H_5CH_2CO_2CH_3 + H_2O$

18.39 **Step 1:** Nitration of the aromatic ring using HNO_3/H_2SO_4 (Section 13.3B). If there is a mixture of nitration products, assume that the desired 4-nitrobenzoic acid can be separated and purified.
Step 2: Catalytic reduction of the nitro group to an amino group (Section 13.3B).

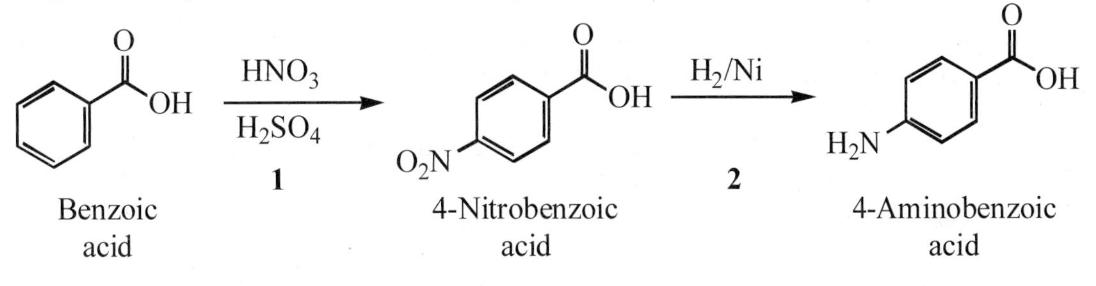

18.41 Each starting material is difunctional and each functional group can participate in formation of an ester, thus giving rise to a polyester.

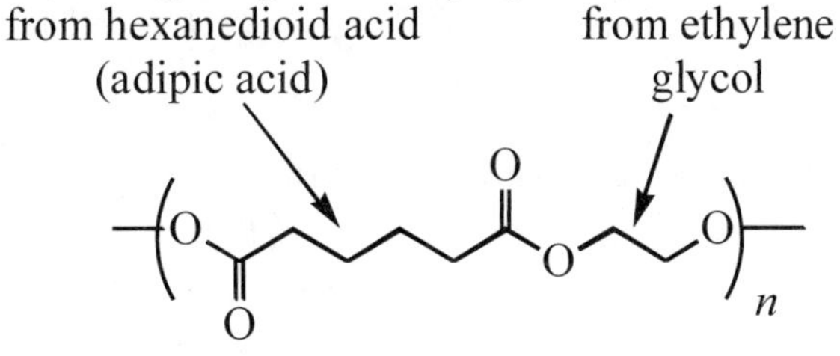

18.43 Following are structural formulas for the product of each oxidation.

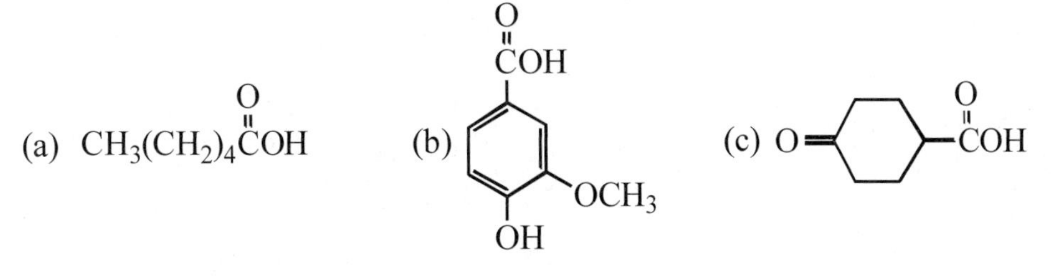

19.1 The following is a structural formula for each amide.

(a) cyclohexyl–$NHCCH_3$ (C=O) (b) phenyl–CNH_2 (C=O)

19.3 In aqueous NaOH, each carboxyl group is present as a carboxylic anion, and each amine is present in its unprotonated form.

(a) $CH_3CN(CH_3)_2 + NaOH \xrightarrow[\text{heat}]{H_2O} CH_3CO^-Na^+ + (CH_3)_2NH$

(b) (six-membered lactam, C=O, NH) $+ NaOH \xrightarrow[\text{heat}]{H_2O} H_2N\text{(CH}_2)_4\text{CO}^-Na^+$

19.5 (a) Benzoic anhydride (b) Methyl decanoate (c) *N*-Methylhexanamide
(d) 4-Aminobenzamide (e) Cyclopentyl acetate (f) Ethyl 3-hydroxybutanoate

19.7 Each reaction brings about hydrolysis of the amide bond. Each product is shown as it would exist under the specified reaction conditions.

(a) $C_6H_5CNH_2 + NaOH \xrightarrow{H_2O} C_6H_5CO^- Na^+ + NH_3$

(b) $C_6H_5CNH_2 + HCl \xrightarrow{H_2O} C_6H_5COH + NH_4^+ Cl^-$

19.9 The products in each part are an amide and acetic acid.

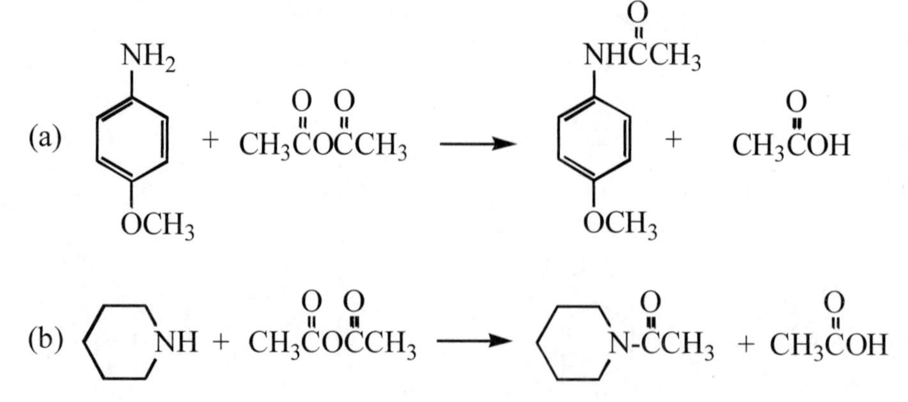

19.11 (a) Phenobarbital contains four amide groups.
(b) Complete hydrolysis of all amide bonds gives a dicarboxylic acid dianion, two moles of ammonia, and one mole of sodium carbonate.

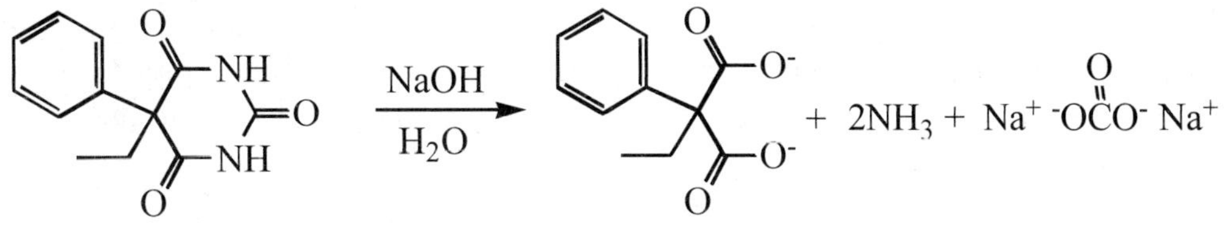

19.13 In nylon-66 and Kevlar, the monomer units are joined by amide bonds.

19.15 In Dacron and Mylar, the monomer units are joined by ester bonds.

19.17 Following are structural formulas for the mono-, di-, and triethyl esters.

$HO-P(=O)(OH)-OCH_2CH_3$	$HO-P(=O)(OCH_2CH_3)-OCH_2CH_3$	$CH_3CH_2O-P(=O)(OCH_2CH_3)-OCH_2CH_3$
Ethyl phosphate	Diethyl phosphate	Triethyl phosphate

19.19 Two molecules of water are split out.

$$HO\text{-}P(=O)(OH)\text{-}OH + HO\text{-}P(=O)(OH)\text{-}OH + HO\text{-}P(=O)(OH)\text{-}OH \longrightarrow HO\text{-}P(=O)(OH)\text{-}O\text{-}P(=O)(OH)\text{-}O\text{-}P(=O)(OH)\text{-}OH + 2H_2O$$

19.21 The arrow points to the ester group. On the right is chrysanthemic acid.

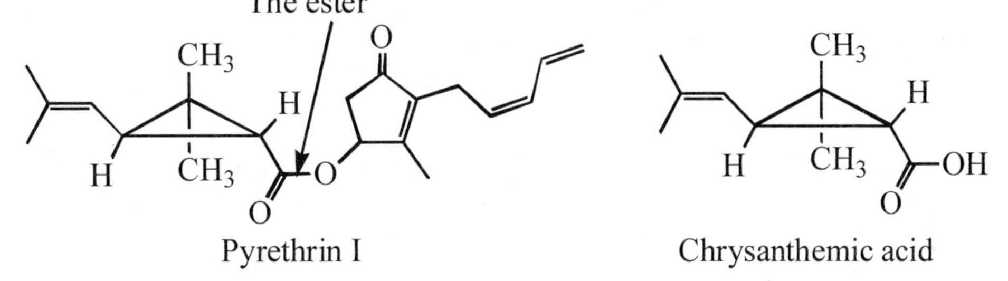

Pyrethrin I

Chrysanthemic acid

19.23 (a) The *cis/trans ratio* refers to the *cis-trans* relationship between the ester group and the carbon-carbon double bond in the three-membered ring.
(b) Permethrin has three stereocenters, and eight stereoisomers (four pairs of enantiomers) are possible for it. The designation "(+/-)" refers to the fact that the members of each pair of possible enantiomers are present in equal amounts; that is, each pair of enantiomers is present as a racemic mixture.

19.25 The compound is salicin. Removal of the glucose unit and oxidation of the primary alcohol to a carboxylic acid gives salicylic acid.

19.27 The moisture present in humid air may be sufficient to bring about hydrolysis of the ester to yield salicylic acid and acetic acid. The vinegar-like odor is due to the presence of acetic acid.

19.29 A *sunblock* prevents all ultraviolet radiation from reaching protected skin by reflecting it away from the skin. A *sunscreen* prevents a portion of the ultraviolet radiation from reaching protected skin. Its effectiveness is related to its skin protection factor (SPF).

19.31 They all contain an ester bonded to an alkyl group as well as a benzene ring. The benzene has either a nitrogen atom or an oxygen atom on it.

19.33 Lactomer stitches dissolve as the ester groups in the polymer chain are hydrolyzed until only glycolic acid and lactic acid remain. These small molecules are metabolized and excreted by existing biochemical pathways.

19.35 Following is an equation for this synthesis of acetaminophen.

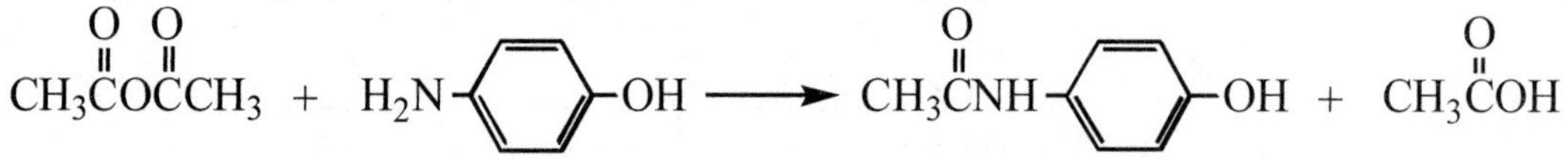

$$CH_3\overset{O}{\overset{\|}{C}}O\overset{O}{\overset{\|}{C}}CH_3 + H_2N\text{-}C_6H_4\text{-}OH \longrightarrow CH_3\overset{O}{\overset{\|}{C}}NH\text{-}C_6H_4\text{-}OH + CH_3\overset{O}{\overset{\|}{C}}OH$$

Acetic anhydride 4-Aminophenol Acetaminophen

19.37 Hydrolysis gives one mole of 2,3-dihydroxypropanoic acid and two moles of phosphoric acid. At pH 7.35 - 7.45, the carboxyl group is present as its anion, and phosphoric acid is present as its dianion.

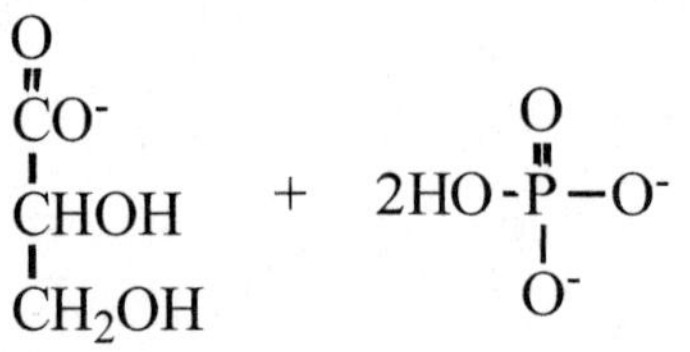

$$\begin{array}{c} O \\ \| \\ CO^- \\ | \\ CHOH \\ | \\ CH_2OH \end{array} + 2HO\text{-}\overset{O}{\overset{\|}{\underset{O^-}{\underset{|}{P}}}}-O^-$$

19.39 Following is a short section of the polyamide formed in this type of polymerization. In the formation of the polymer, a molecule of water is lost for every amide bond formed.

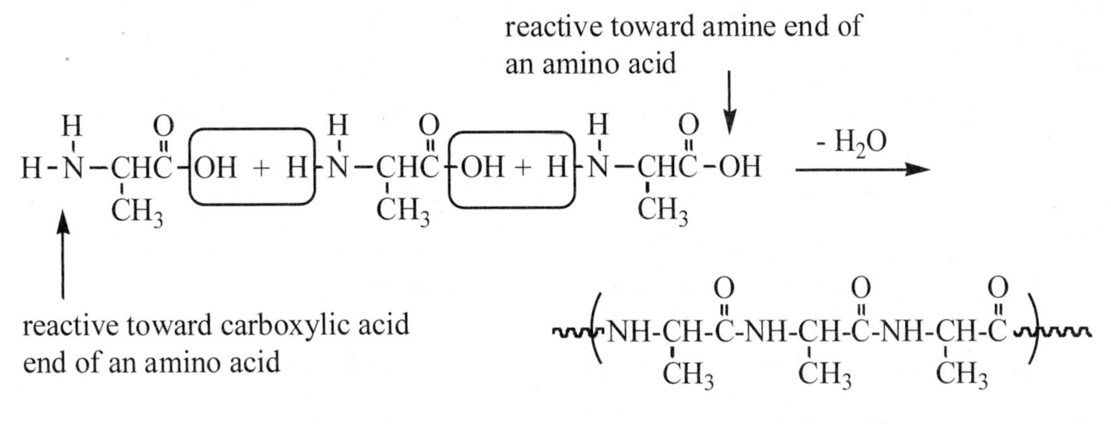

19.41 Following are structural formulas for the reagents to synthesize each amide.

(a) $-NH_2$ + $HO\overset{O}{\overset{\|}{C}}(CH_2)_4CH_3$ (b) $(CH_3)_2CH\overset{O}{\overset{\|}{C}}OH$ + $HN(CH_3)_2$

(c) $2NH_3$ + $HO\overset{O}{\overset{\|}{C}}(CH_2)_4\overset{O}{\overset{\|}{C}}OH$

19.43 (a) Both lidocaine and mepivicaine contain an amide group and a tertiary aliphatic amine.

(b) Both local anesthetics share the structural features shown enclosed in the box at the right.

H N N O

Lidocaine (Xylocaine)

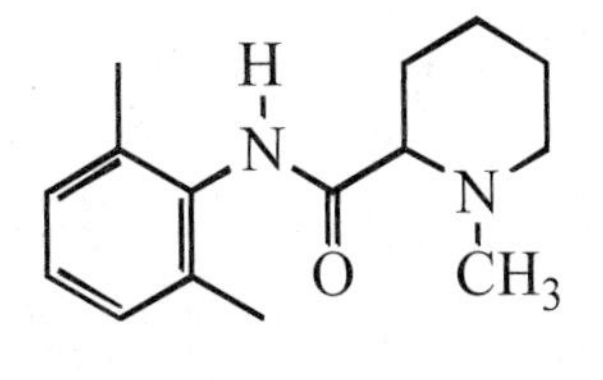

Mepivacaine (Carbocaine)

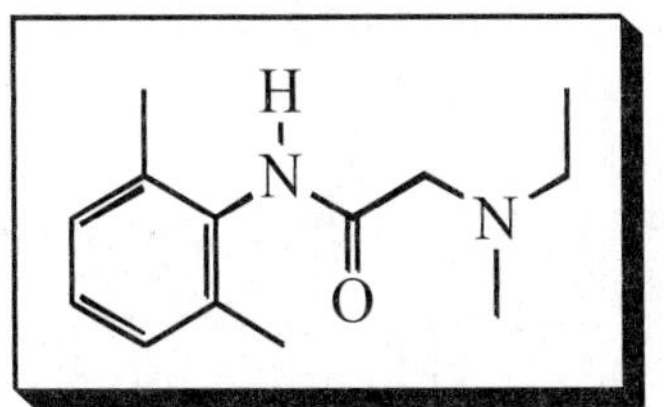

Structural features common to both local anesthetics

20.1 Following are Fischer projections for the four 2-ketopentoses. They consist of two pairs of enantiomers.

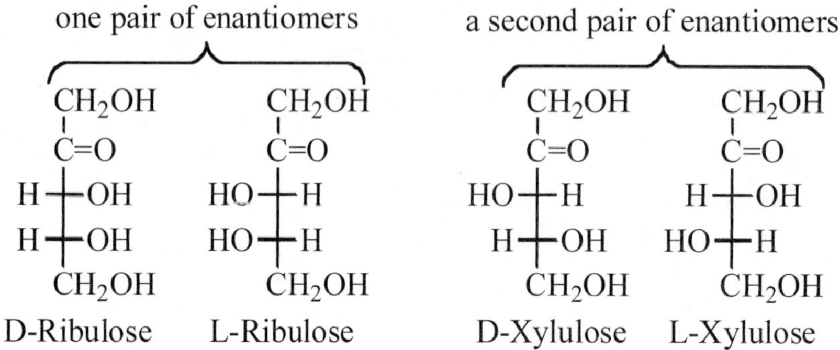

20.3 D-Mannose differs in configuration from D-glucose only at carbon 2.

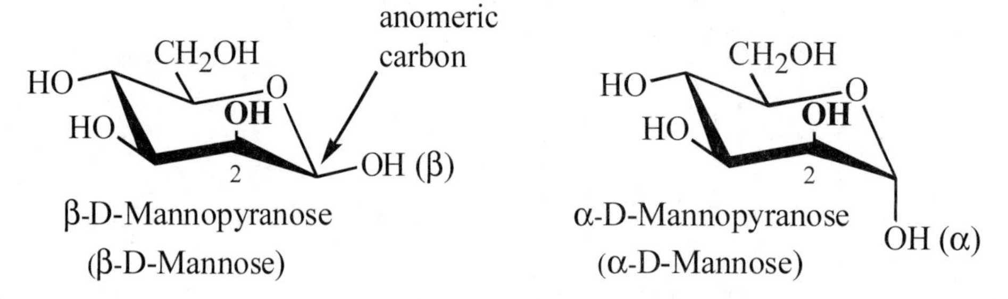

20.5 The β-glycosidic bond is between carbon 1 of the left unit and carbon 3 of the right unit.

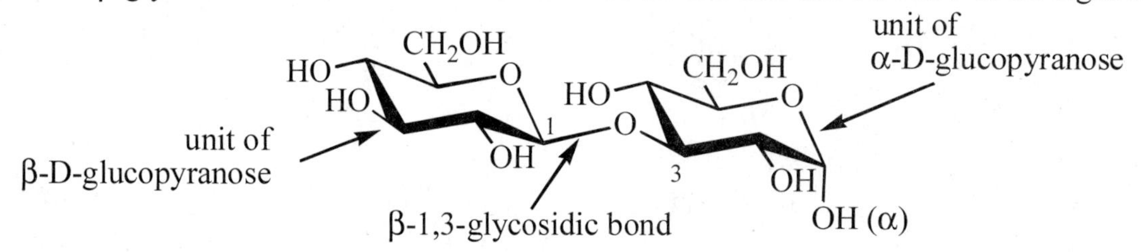

20.7 The carbonyl group in an aldose is an aldehyde. In a ketose, the carbonyl is a ketone.

20.9 The three most abundant hexoses are D-glucose, D-galactose, and D-fructose. The first two are aldohexoses. The third is a 2-ketohexose.

20.11 To say that they are enantiomers means that they are nonsuperposable mirror images.

20.13 The D or L configuration in an aldopentose is determined by its configuration at carbon 4.

20.15 Compounds (a) and (c) are D-monosaccharides. Compound (b) is an L-monosaccharide.

20.17 A 2-ketoheptose has four stereocenters and 16 possible stereoisomers. Eight of these are D-2-ketoheptoses and eight are L-2-ketoheptoses. Following is one of the eight possible D-2-ketoheptoses.

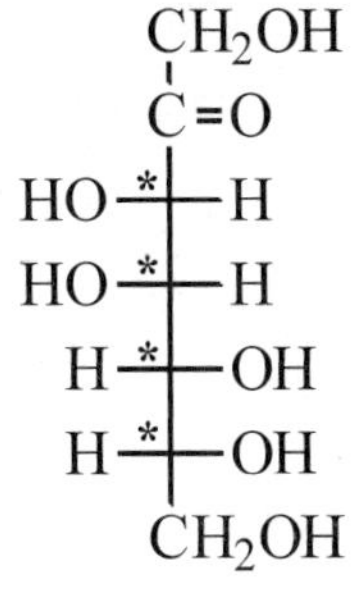

20.19 In an amino sugar, one or more –OH groups are replaced by an $-NH_2$ groups. The three most abundant amino sugars in the biological world are D-glucosamine, D-galactosamine, and *N*-acetyl-D-glucosamine.

20.21 (a) A pyranose is a six-membered cyclic hemiacetal form of a monosaccharide.
(b) A furanose is a five-membered cyclic hemiacetal form of a monosaccharide.

20.23 Yes, they are anomers. No, they are not enantiomers; that is, they are not mirror images. They differ in configuration only at carbon 1 and, therefore, they are diastereomers.

20.25 A Haworth projection shows the six-membered ring as a planar hexagon. In reality, the ring is puckered and its most stable conformation is a chair conformation with all angles approximately 109.5°.

20.27 One trick to solving this problem is to first identify the difference between compounds (a) and (b) and D-glucose. In the chair conformation, β-D-glucopyranose places all of the –OH and $-CH_2OH$ groups in the equatorial position. Ignoring the configuration at the anomeric carbon (C1) in the cyclic form, compound (a) differs from D-glucose only in the configuration at carbon 4. Compound (b) differs only at carbon 3.

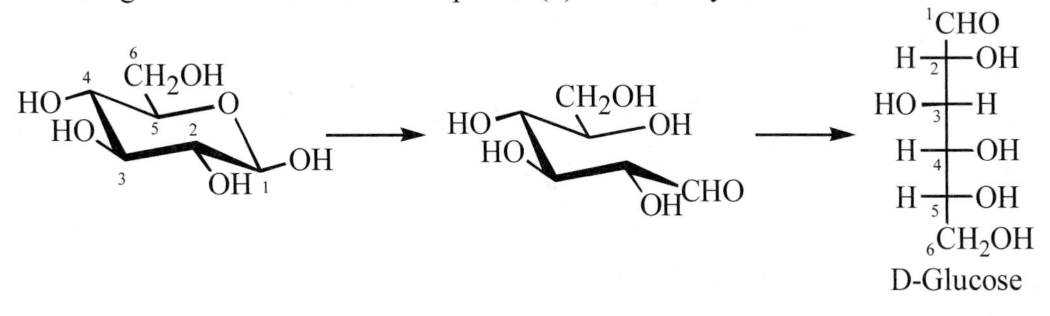

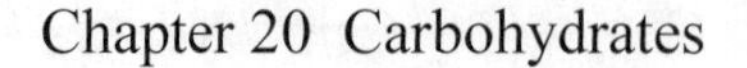

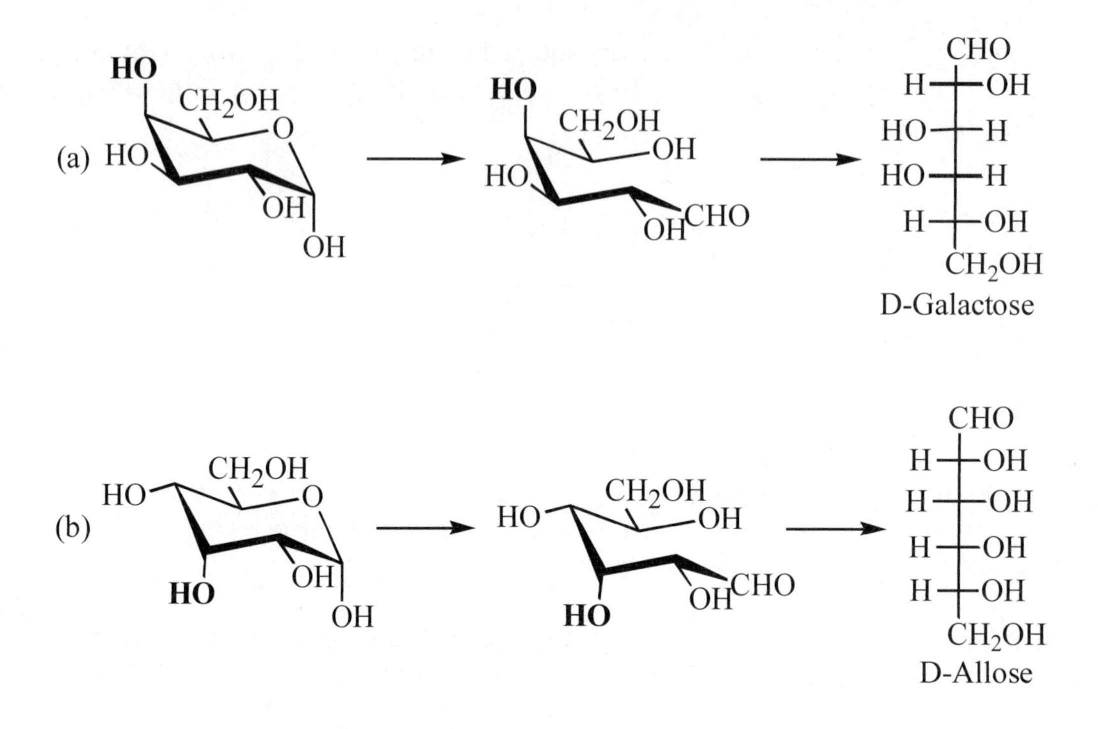

20.29 The specific rotation of α-L-glucose is -112.2°.

20.31 A glycoside is a cyclic acetal of a monosaccharide. A glycosidic bond is the bond from the anomeric carbon to the –OR group of the glycoside.

20.33 No, glycosides cannot undergo mutarotation because the anomeric carbon is not free to interconvert between α and β configurations via the open-chain aldehyde or ketone.

20.35 Following are Fischer projections of D-glucose and D-sorbitol. The configurations at the four stereocenters of D-glucose are not affected by this reduction.

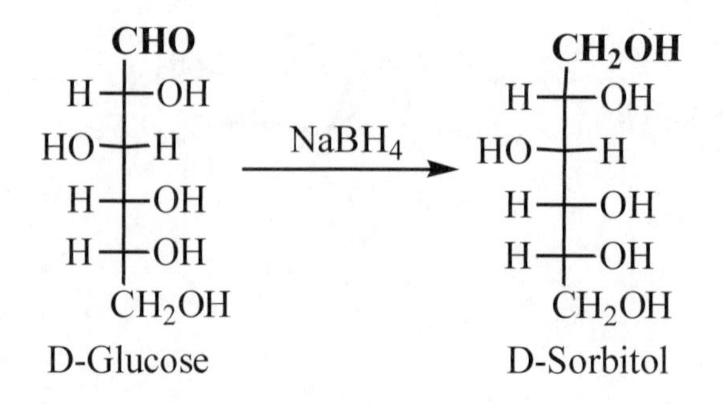

20.37 Ribitol is the reduction product of D-Ribose. β-D-Ribose 1-phosphate is the phosphoric ester of the –OH group on the anomeric carbon of β-D-ribofuranose.

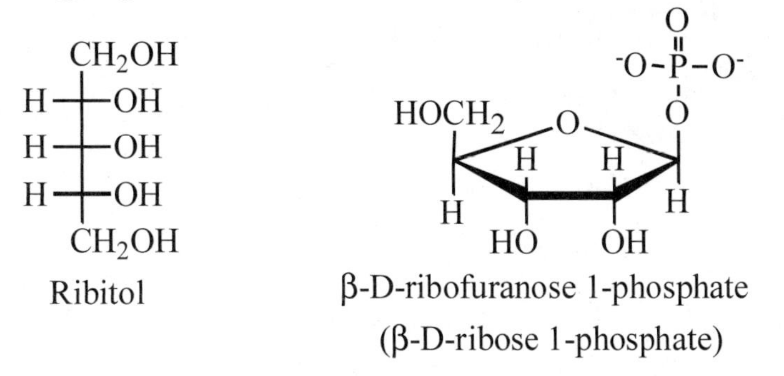

20.39 A β-1,4-glycosidic bond indicates that the configuration at the anomeric carbon (carbon 1 in this problem) of the monosaccharide unit forming the glycosidic bond is beta and that it is bonded to carbon 4 of the second monosaccharide unit. An α-1,6-glycosidic bond indicates that the configuration at the anomeric carbon (carbon 1 in this problem) of the monosaccharide unit forming the glycosidic bond is alpha and that it is bonded to carbon 6 of the second monosaccharide unit.

20.41 (a) Both monosaccharide units are D-glucose.
(b) They are joined by a β-1,4-glycosidic bond.
(c) It is a reducing sugar.
(d) It undergoes mutarotation.

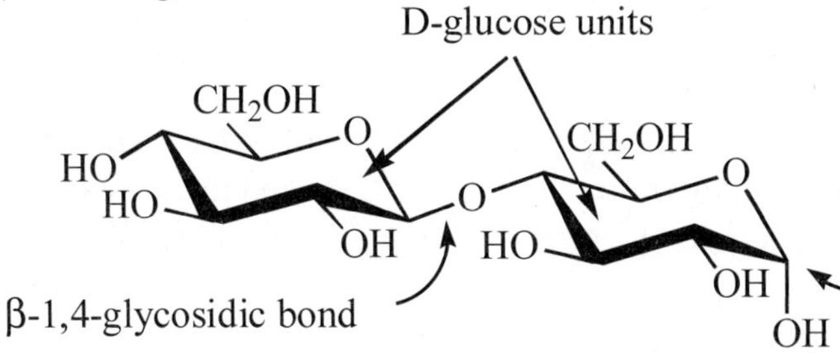

20.43 An oligosaccharide contains approximately four to ten monosaccharide units. A polysaccharide contains more, generally many more, than ten monosaccharide units.

20.45 The difference is in the degree of chain branching. Amylose is composed of unbranched chains, whereas amylopectin is a branched network with branches started by β-1,6-glycosidic bonds.

20.47 Cellulose fibers are insoluble in water because the strength of hydrogen bonding of a cellulose molecule in the fiber with surface water molecules is not sufficient to overcome the intermolecular forces that hold it in the fiber.

20.49 (a) In the following structural formula, the $CH_3C{=}O$ (the acetyl group) is abbreviated Ac.

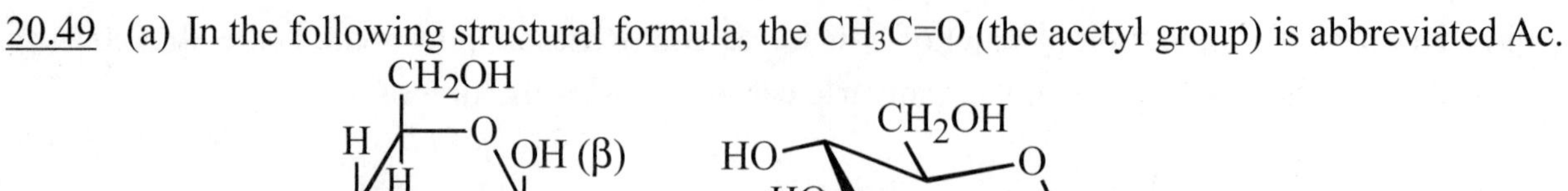

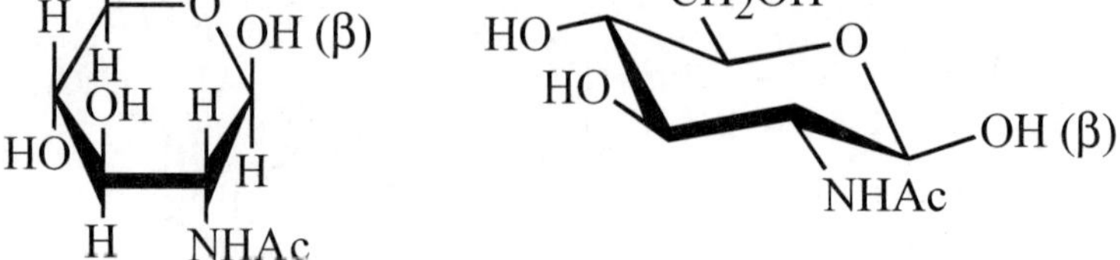

(b) The following Haworth and chair structures for this repeating disaccharide.

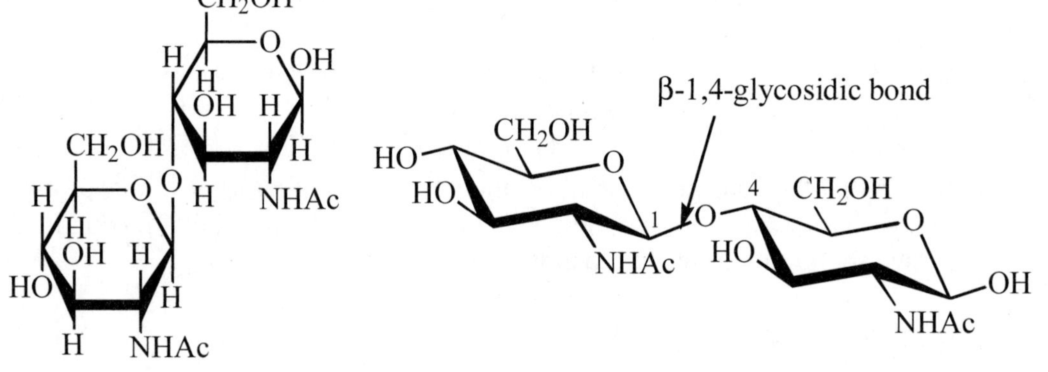

20.51 Its lubricating power decreases.

20.53 Sucrose is a disaccharide composed of D-glucose and D-fructose, thus lacks the problematic D-galactose. With maturation, children develop an enzyme capable of metabolizing D-galactose; therefore, they are able to tolerate galactose as they get older.

20.55 L-Ascorbic acid is oxidized (there is loss of two hydrogen atoms) when it is converted to L-dehydroascorbic acid. L-Ascorbic acid is a biological reducing agent.

20.57 Types A, B and O have in common D-galactose and L-fucose. Only type A has N-acetyl-D-glucosamine.

20.59 Mixing types A and B blood will result in coagulation.

20.61 Consult Table 20.1 for the structural formula of D-altrose and draw it. Then replace the -–OH groups on carbons 2 and 6 by hydrogens.

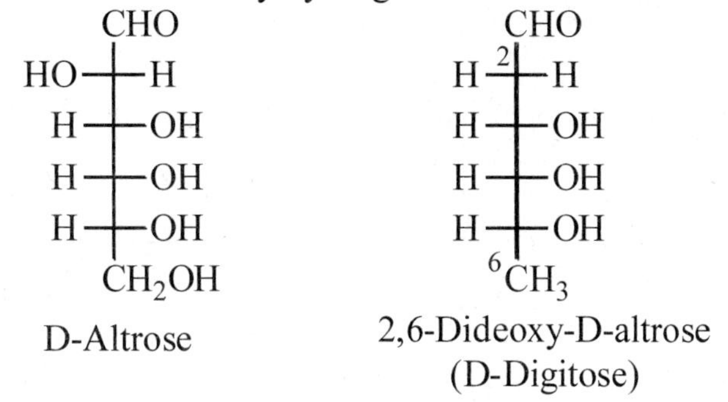

20.63 The monosaccharide unit in salicin is D-glucose.

20.65 Review Section 17.5 on keto-enol tautomerism and your answer to Problem 17.64. The intermediate in this conversion is an enediol; that is, it contains a carbon-carbon double bond with two OH groups on it.

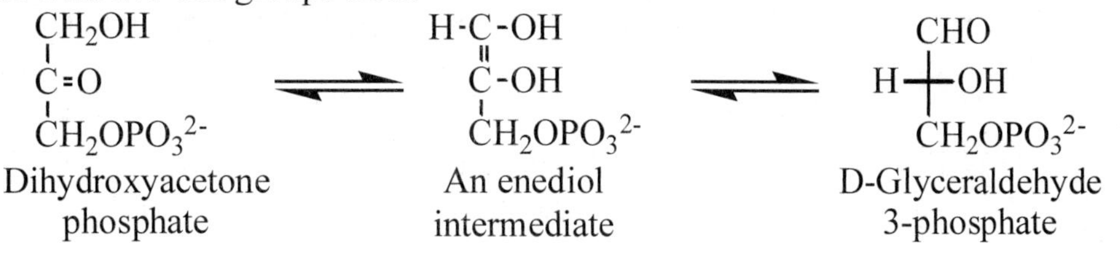

20.67 (a) Coenzyme A is chiral. It has five stereocenters, which are identified with an asterisk.

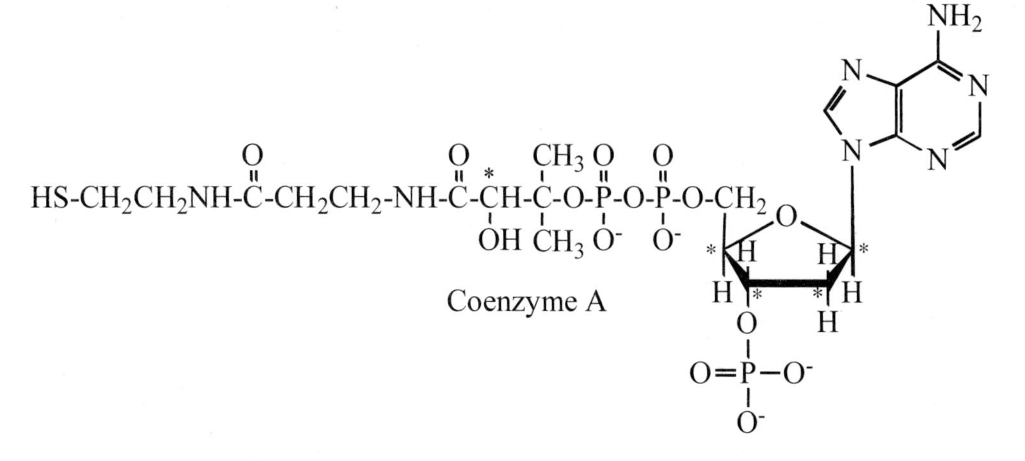

(b) The functional groups, starting from the left, are a thiol (-SH), two amides, a secondary alcohol, a phosphate ester, a phosphate anhydride, a phosphate ester, another phosphate ester, a unit of 2-deoxyribose, and a β-glycosidic bond to adenine, a heterocyclic amine.

(c) Yes, it is soluble in water because of the presence of a number of polar C=O groups, one –OH group and three phosphate groups, all of which will interact with water molecules by hydrogen bonding.

(d) Following are the products of hydrolysis in aqueous HCl of all amide, ester and glycosidic bonds.

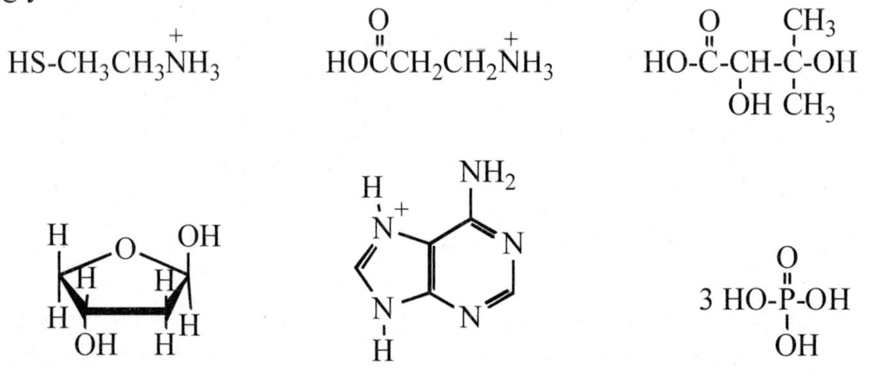

(e) Following are the products of hydrolysis in aqueous NaOH of all amide, ester, and glycosidic bonds.

$^{-}S\text{-}CH_2CH_2NH_2$ $\qquad$ $^{-}OC(=O)CH_2CH_2NH_2$ $\qquad$ $^{-}O\text{-}C(=O)\text{-}CH(OH)\text{-}C(CH_3)_2\text{-}OH$

$3\ ^{-}O\text{-}P(=O)(O^{-})\text{-}O^{-}$

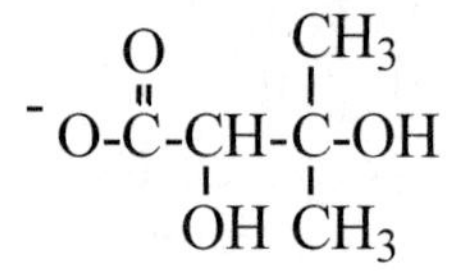

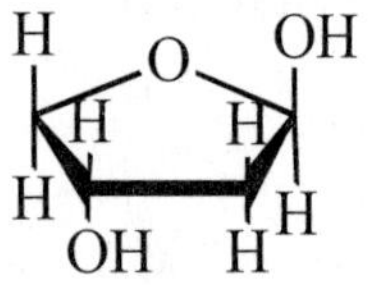

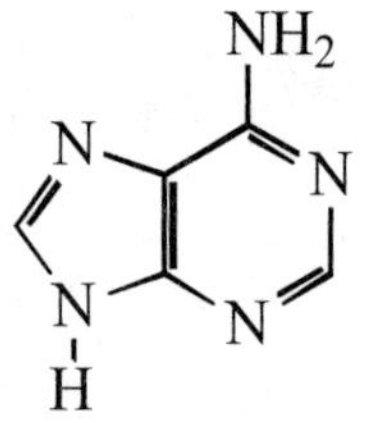

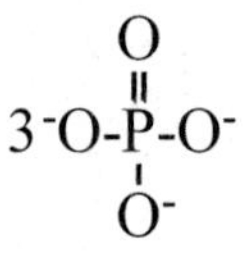

Chapter 21 Lipids

21.1 (a) This complex lipid is produced from the triol glycerol. Two of the hydroxyl groups are esterified with fatty acids, the third with phosphoric acid. Therefore, the lipid is broadly classified as a glycerophospholipid. The phosphate group is esterified with the hydroxyl group of the amino acid serine; therefore, the lipid belongs to the subgroup cephalins.
(b) The components present: glycerol backbone, myristic acid, linoleic acid, phosphate, and serine.

21.3 The term hydrophobic means water fearing. Hydrophobic molecules, also called nonpolar molecules, are usually not soluble in water, but they are soluble in nonpolar organic solvents such as pentane or dichloromethane. The hydrophobic nature of lipids is important because for cells to function properly, they require separation of aqueous components (organelles). These components are surrounded by membranes that are composed of hydrophobic lipids and thus do not allow free flow of water and many other cellular molecules. Biochemical processes may be separated this way.

21.5 The melting points of fatty acids (Tables 18.3 and 21.1) are dependent on the length of the carbon chains and the number and types of double bonds. Cis double bonds cause a kink in the chain (see Section 21.2) that disrupts the London dispersion forces between the chains and lowers the melting point of oleic acid (18:1) compared to the saturated acid, stearic acid (18:0). However, the trans fatty acid (18:1), resembles a saturated acid and does not disrupt the chain as much as the cis acid. Thus the melting point would be predicted to be higher. Actual melting points: 18:0 (70° C); 18:1 trans (45° C); 18:1 cis (16° C).

21.7 The diglycerides with the highest melting points will be those with two stearic acids (a saturated fatty acid). The lowest melting point will be the one with two oleic acids (a monounsaturated fatty acid).

21.9 A triglyceride containing only stearic acid (18:0) will have a higher melting point than a triglyceride containing only lauric acid (12:0). The correct answer is (b).

21.11 The highest melting triglyceride would be (a), containing palmitic and stearic (both saturated fatty acids). The lowest melting triglyceride is (c) because all fatty acids are unsaturated. Triglyceride (b), which has one unsaturated fatty acid and two saturated acids, would have a melting point in between triglycerides (a) and (c). Therefore, the order from lowest to highest is (c), (b), (a)

21.13 In order of increasing solubility in water, they are (a) < (b) < (c). Solubility in water is dependent on the presence of polar functional groups like the hydroxyl group. The greater the number of hydroxyl groups, the greater the solubility. A monoglyceride has two hydroxyl groups, a diglyceride has one hydroxyl group, and a triglyceride has none. Fatty acids are essentially insoluble in water.

21.15 Saponification of the triglyceride hydrolyzes the ester bonds yielding glycerol and the sodium salts of the fatty acids palmitic, stearic, and linolenic.

21.17 Complex lipids can be divided into two groups: phospholipids and glycolipids. Phospholipids contain an alcohol, two fatty acids, and a phosphate group. There are two types: glycerophospholipids and sphingolipids. In glycerophospholipids, the alcohol is glycerol. In sphingolipids, the alcohol is sphingosine. Glycolipids are complex lipids that contain carbohydrates.

21.19 The fluidity of a membrane is dependent on the amount of cholesterol and on the amounts and types of fatty acids in the phospholipid bilayer. The presence of cis double bonds causes greater fluidity because they cannot pack together as closely as saturated fatty acids. The greater the concentration of lipids with unsaturated fatty acids, the greater the fluidity of the membrane.

21.21 Integral membrane proteins are embedded within the membrane and are held in position by very strong hydrophobic interactions. They often span the entire lipid bilayer where they serve as channels through which molecules are transported in and out of a cell. Peripheral membrane proteins are present on the surfaces of the membrane where they are weakly attached.

21.23 A phosphatidylinositol containing oleic acid and arachidonic acid:

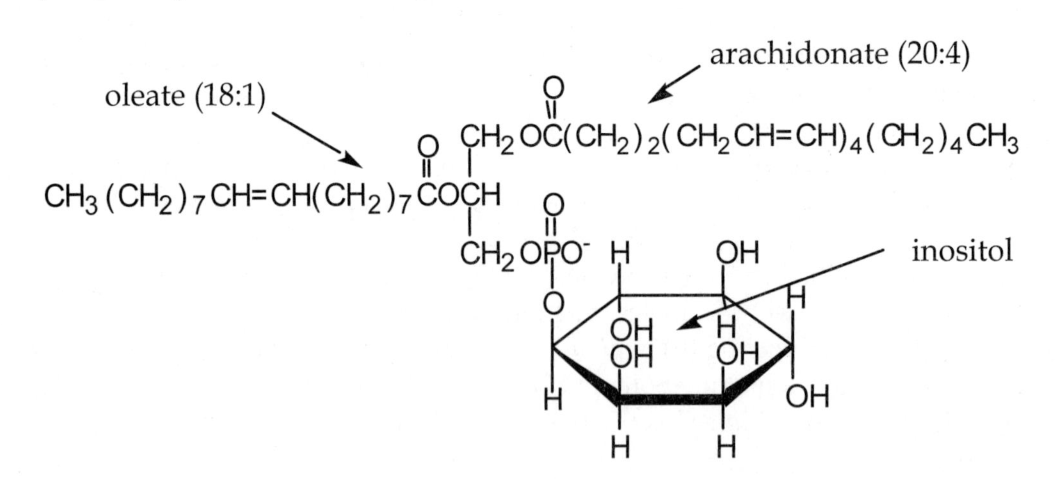

21.25 Complex lipids that contain ceramides include sphingomyelin, a sphingolipid, and the cerebroside glycolipids.

21.27 The hydrophilic functional groups of (a) glucocerebroside: carbohydrate; hydroxyl and amide group of the ceramide.
(b) sphingomyelin: phosphate group; choline; hydroxyl and amide of ceramide.

21.29 Cholesterol crystals may be found in (1) gallstones, that are sometimes pure cholesterol; (2) joints of people suffering from bursitis.

21.31 Carbon 17 of the D steroid ring undergoes the most substitution.

21.33 LDL in the bloodstream is delivered to cells by binding to LDL receptor proteins in areas called coated pits on the surfaces of the cells. After binding, the LDL is transported inside the cells (endocytosis) where the cholesterol is released by enzymatic degradation of the LDL.

21.35 VLDL particles, which have mainly a triglyceride core, are produced in the liver. They enter the blood and are carried to fat and muscle tissue where they deliver triglycerides for energy metabolism. Removing the lipid from the VLDL core increases their density and converts them to LDL particles.

21.37 Serum cholesterol levels control the production of LDL receptors that help cells take up cholesterol in the form of LDL. When serum cholesterol concentration is high, the synthesis of cholesterol in the liver is inhibited and the synthesis of cell receptors is increased. Serum cholesterol levels control the formation of cholesterol in the liver by regulating enzymes that synthesize cholesterol. This is an example of feed-back control.

21.39 Estradiol (E) is synthesized from progesterone (P) through the intermediate testosterone (T). First the D-ring acetyl group of P is converted to a hydroxyl group and T is produced. The methyl group in T, at the junction of rings A and B, is removed and ring A becomes aromatic. The keto group in P and T is converted to a hydroxyl group in E.

21.41 The structures of the three steroids are shown in Figure 21.6. They have similar substituents at several positions, but differ greatly at C-11. Progesterone has no substituents except hydrogen, cortisol has a hydroxyl group, cortisone has a keto group, and RU486 has a large *p*-aminophenyl group. Apparently the functional group at C-11 is of little importance in binding the compounds to the receptor.

21.43 Common structural features in oral contraceptives: they all are based on the four-fused steroid ring; most have structures similar to progesterone; all have a methyl group at C-13; all have an acetylenic group on C-17; all have some unsaturation in ring A and/or B.

21.45 The bile salts help solubilize fats during the digestion process and elimination of waste products. They assist in removal of excess cholesterol that causes plaque formation in atherosclerosis in two ways: (1) They themselves are oxidation products of cholesterol degradation so this decreases cholesterol concentrations, and (2) they bind to cholesterol and form complexes that are eliminated in the feces.

21.47 (a) Glycocholate:

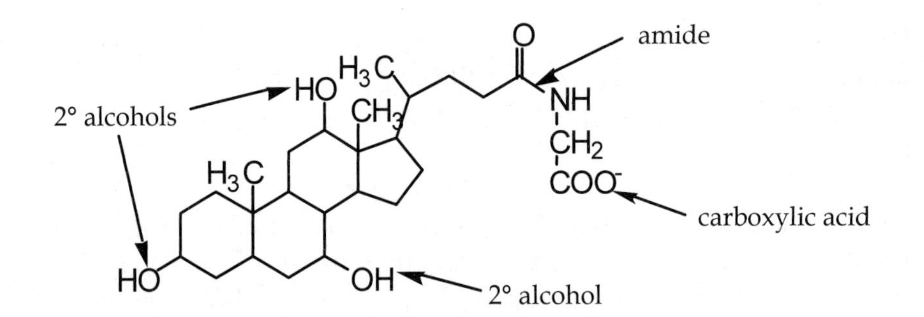

(b) Cortisone:

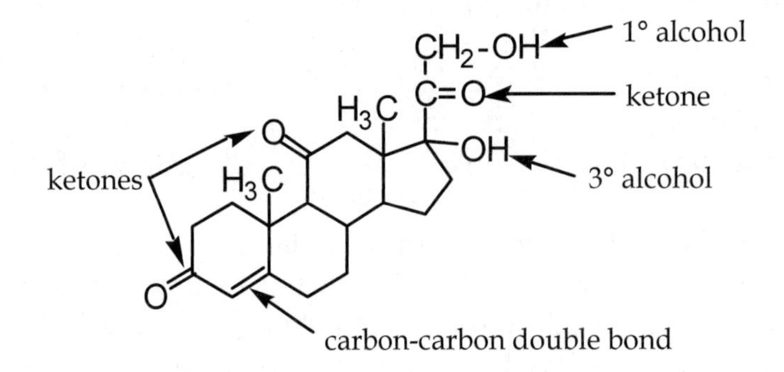

(c) PGE_2:

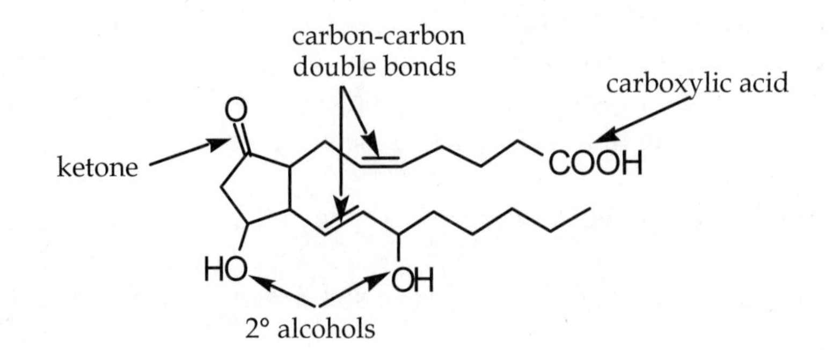

(d) Leukotriene B4:

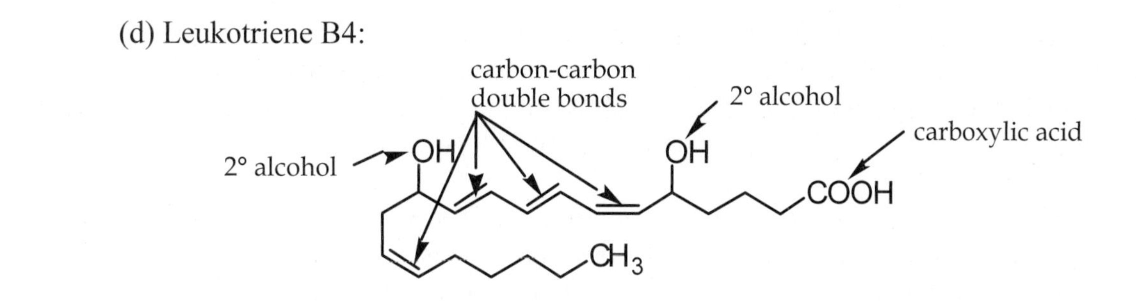

21.49 Thromboxanes, which are derived from arachidonic acid via the prostaglandins, stimulate platelet aggregation thus enhancing the blood clotting process. Aspirin slows the synthesis of the thromboxanes by inhibiting the COX enzyme. Thus in the presence of aspirin, there will be lowered amounts of thromboxanes and the blood clotting process will be slowed. Strokes caused by blood clots in the brain will occur less often.

21.51 The major components of waxes are esters of long chain acids and alcohols. Beeswax contains palmitic acid (16:0) and 1-triacontanol, a saturated, unbranched, primary alcohol with thirty carbons. Because of the presence of saturated components, wax molecules pack closer together in the solid form and have higher melting points than triglycerides that often contain many unsaturated components.

21.53 The anion transporter of the erythrocyte membrane exchanges chloride and bicarbonate ions. The protein transporter forms a hydrophilic channel through the membrane. Protein polar groups in the channel wall facilitate the transport of hydrated chloride and bicarbonate ions. The hydrophobic groups on the outer surfaces of the helices interact with the membrane.

21.55 (a) Nerve axons are surrounded by a lipid layer, called the myelin sheath, which is composed primarily of sphingomyelin. The protective lipid layer provides insulation and enhances the rapid conduction of electrical signals.
(b) In multiple sclerosis, there is slow degradation of the protective myelin sheath. This exposes the nerve axons to damage and causes afflicted individuals to lose strength and coordination.

21.57 The glycolipid that accumulates in Fabry's disease contains the monosaccharides galactose and glucose.

21.59 Progesterone is secreted during the menstrual cycle after ovulation (see Figure 21.7). Its purpose is to prevent another ovulation if the egg is fertilized. Contraceptive agents give the user a level of progesterone that sends the message that there is a fertilized egg; therefore, ovulation does not occur.

21.61 Indomethacin, a non-steroidal anti-inflammatory agent, acts by inhibiting the cyclo-oxygenase enzymes that catalyze the formation of prostaglandins from arachidonic acid. Prostaglandins cause an inflammatory response so inhibiting their formation will reduce inflammation.

21.63 The leukotrienes are synthesized from arachidonic acid as are prostaglandins, but not through the same pathway using the COX enzymes. Thus, the NSAIDS inhibit the cyclo-oxygenases and prostaglandin synthesis, but do not effect leukotriene synthesis. Aspirin and other NSAIDS agents can reduce inflammation, but they do not alter the physiological actions of the leukotrienes.

21.65 (Study membrane structure in Figure 21.2). Three regions of the membrane need to be considered in order to define transport: 1) the cytoplasmic surfaces and 2) the exterior surfaces that are comprised primarily of the polar heads of complex lipids (purple balls) and 3) the membrane interior composed of the hydrophobic tails of the lipids. To diffuse, molecules must pass through all three regions, two polar regions and one nonpolar region. Polar molecules interact favorably with the surface regions by hydrogen bonding and ion-dipole interactions, but they repel the hydrophobic barrier, thus they would not diffuse through the membrane. In addition, polar molecules would prefer to interact with water that is present at each surface. On the other hand, small, nonpolar molecules are able to diffuse because they slip through surface regions containing mainly hydrophobic components (lipid patches).

21.67 Although prostaglandins and thromboxanes have different structures, the physiological production of both is slowed by COX inhibitors. This suggests that there must be a common step in their syntheses. Prostaglandins are synthesized from arachidonic acid using the COX enzymes. Thromboxanes are synthesized from the prostaglandin PGH.

21.69 Coated pits are areas on a cell membrane that have high concentrations of LDL receptors and participate in transporting LDLs into the cell.

21.71 In facilitated transport, a membrane protein assists in the movement of a molecule through the membrane. The protein often becomes a channel for transport of the molecule. No energy is required for the transport process. In the active transport of a molecule, a membrane protein assists in the process, but energy, usually released from ATP hydrolysis, is required.

21.73 Aldosterone has an aldehyde group at carbon 13.

21.75 The formula weight of the triglyceride is about 850 g/mole. 100 g of the triglyceride is 100 g ÷ 850 g/mole = 0.12 mole. One mole of hydrogen is required for each mole of double bonds in the triglyceride. There are three double bonds so the moles of hydrogen required for 100 g = 0.12 mole x 3 = 0.36 mole of hydrogen gas. Converting to grams of hydrogen, 0.36 x 2 g/mole = 0.72 g hydrogen gas.

21.77 Both lipids and carbohydrates contain carbon, hydrogen, and oxygen. Carbohydrates have aldehyde and ketone groups, as do some steroids. Carbohydrates have a number of hydroxyl groups, which lipids do not have to a great extent. Lipids have major components that are hydrocarbon in nature. These structural features imply that carbohydrates tend to be significantly more polar than lipids.

21.79 Primarily lipid: olive oil and butter; primarily carbohydrate: cotton and cotton candy.

21.81 The amounts are the key point here. Large amounts of sugar can provide energy. Fat burning due to the presence of taurine plays a relatively minor role because of the small amount.

21.83 The other end of the molecules involved in the ester linkages in lipids, such as fatty acids, tend not to form long chains of bonds with other molecules. Amino acids have two ends with functional groups, so these groups can link together to form chains.

21.85 The bulkier molecules will tend to be found on the exterior of the cell because the curvature of the cell membrane will provide more room for them.

21.87 The charges will tend to cluster on membrane surfaces. Positive and negative charges will attract each other. Two positive or two negative charges will repel, so unlike charges do not have this repulsion.

22.1 The dipeptide Val-Phe:

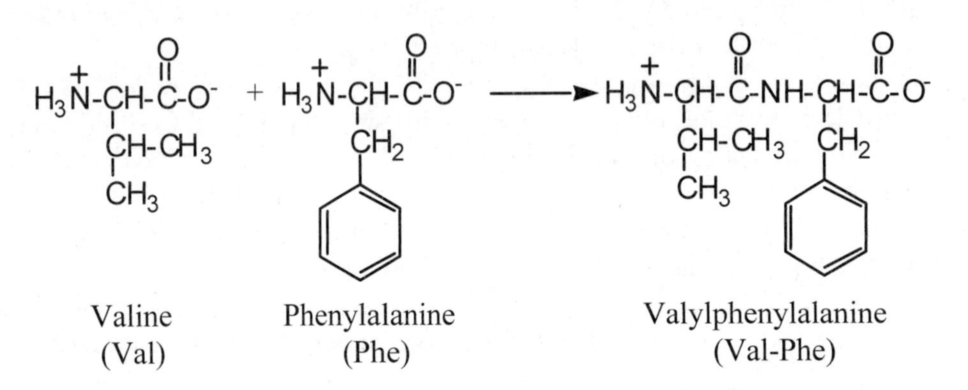

22.3 (a) Ovalbumin is a protein found in eggs that serves to store nutrients for the developing embryo.
(b) The protein myosin is essential for muscle contraction.

22.5 Immunoglobulins are complex protein structures that serve for protection. Antibodies are immunoglobulins, as are certain receptors on T cells and B cells.

22.7 See Figure 22.1. Tyr has a hydroxyl group on the phenyl ring; Phe has a hydrogen at the same position.

22.9 Arg has the highest percentage of nitrogen at 32%. The four nitrogens have a weight of 4 x 14 = 56. The molecular weight of Arg is 174. Percent of N = 56 ÷ 174 x 100 = 32%. His has 27% nitrogen.

22.11 The amino acid Pro, which has a five-membered ring containing a single nitrogen atom, is classified as a pyrrolidine, a heterocyclic aliphatic amine:

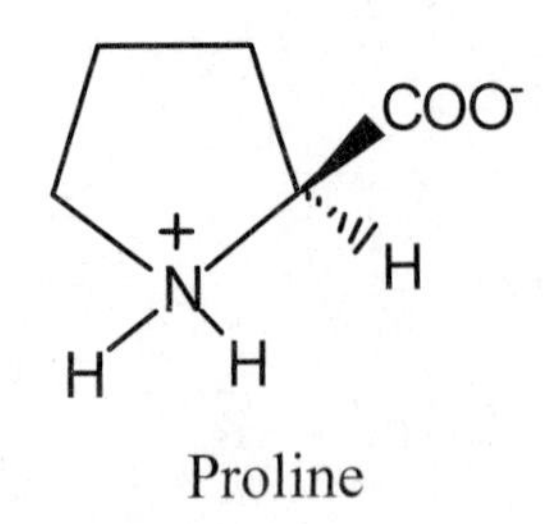

Proline

22.13 Proteins that are ingested from meat, fish, eggs, etc. are degraded to free amino acids and then used to build proteins that carry out specific functions (see Section 22.1). Two very important functions include structural integrity (collagen, keratin) to hold body tissue together, and biological catalysis to assist in cellular reactions, especially metabolism.

22.15 They both have the three-carbon skeleton found in Ala, but Phe has a phenyl group substituted on the β-methyl group:

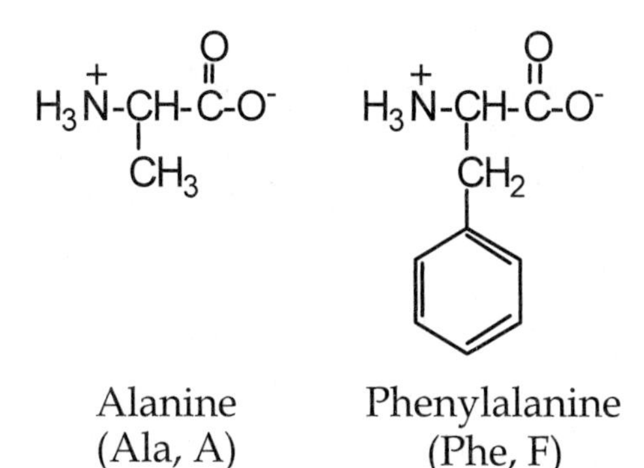

Alanine
(Ala, A)

Phenylalanine
(Phe, F)

22.17 Amino acids exist as zwitterions, compounds that have positive and negative charges on the same molecule. The molecules can interact with each other by strong electrostatic (ionic) bonds. The molecules can pack tightly together into crystalline forms like salts.

22.19 Amino acids have a carboxylic acid group that by nature wants to donate a proton. They also contain a basic amino group that wants to accept a proton. These two groups prefer to react in an acid/base reaction to form a zwitterion:

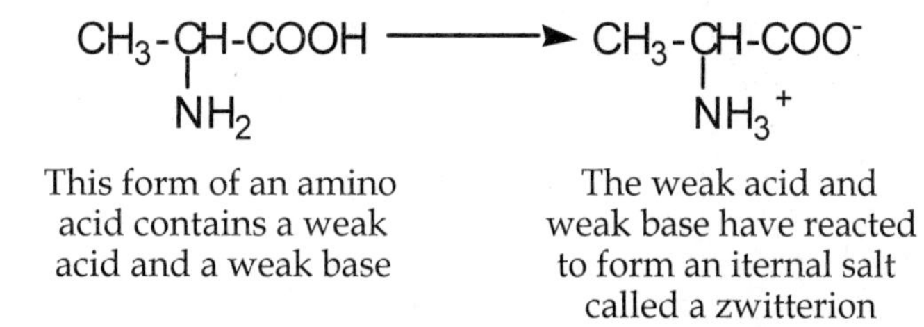

This form of an amino acid contains a weak acid and a weak base

The weak acid and weak base have reacted to form an iternal salt called a zwitterion

22.21 The side chain imidazole is the most unique. All amino acids have an α-carboxyl, and all have an α-amino. Only histidine has a sidechain imidazole, a heterocyclic aromatic amine.

22.23 The side chain of histidine is an imidazole with a nitrogen that reversibly binds to a hydrogen. When dissociated it is neutral; when associated it is positive. Therefore, chemically it is a base, even though it does have a pKa in the acidic range.

22.25 Histidine, Arginine, and Lysine

22.27 Serine may be obtained by the hydroxylation of alanine. Tyrosine is obtained by the hydroxylation of phenylalanine.

22.29

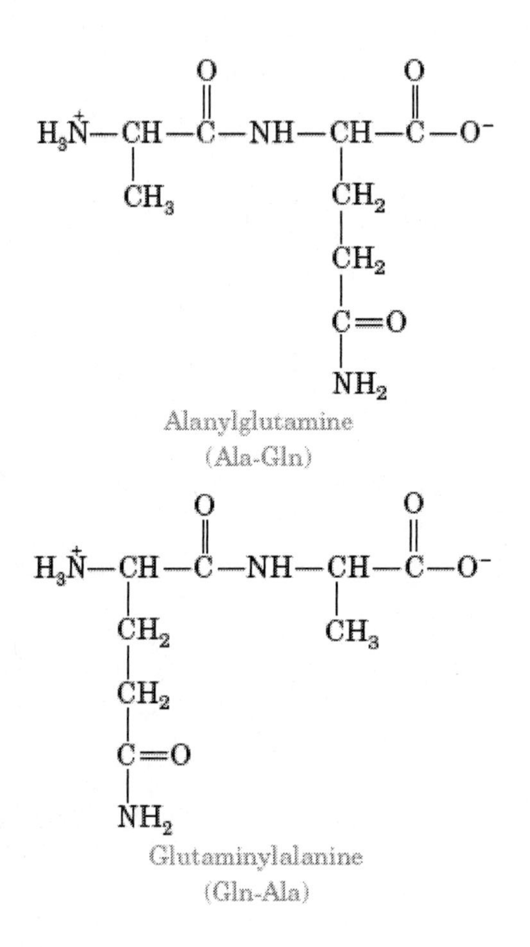

22.31 The tripeptide Thr-Arg-Met:

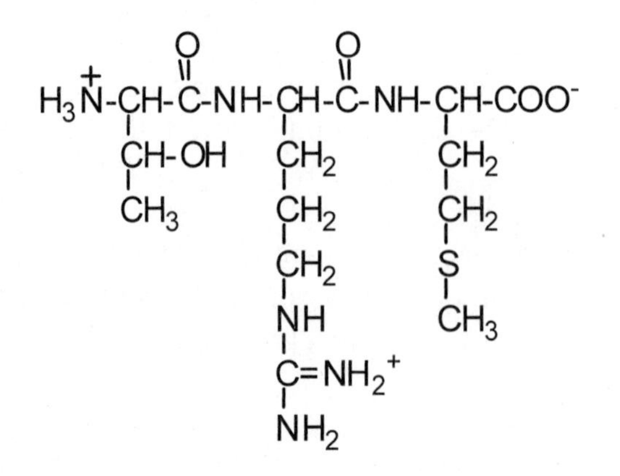

22.33 The side chains of the polypeptide of alternating Val and Phe are very nonpolar (only carbon and hydrogen atoms). The backbone, composed of the repeating pattern of peptidebonds, has atoms that show polarity. The N-H and O = C of the amide bonds may hydrogen bond with each other (as in an α-helix) or these groups may hydrogen bond with water.

22.35 (a) Structure of Met-Ser-Cys:

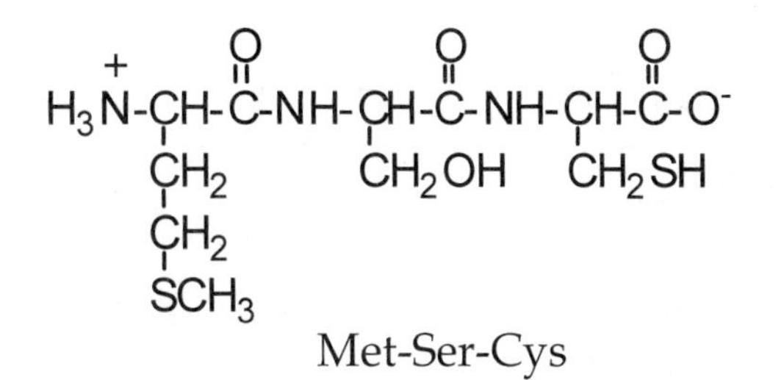

Met-Ser-Cys

(b) The pK_a of the carboxyl group is about 2, so at pH = 2:

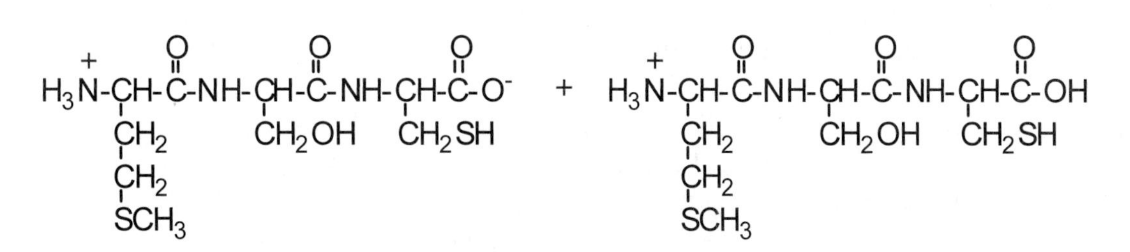

The pK_a of the α-amino group is about 9:

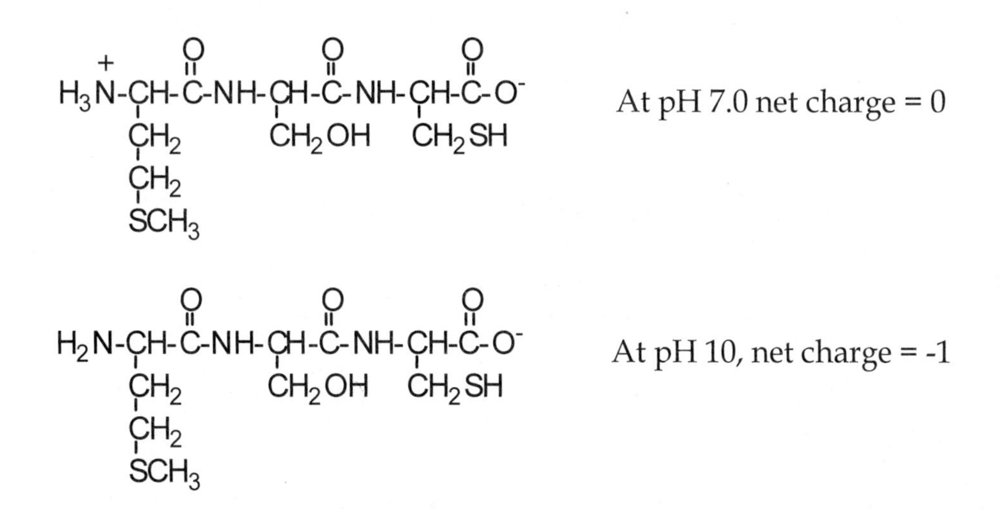

22.37 The precipitated protein at its isoelectric point has an equal number of positive and negative charges (no net charge). The protein molecules clump together and form aggregates. Adding acid protonates the carboxylate groups on amino acid residue side chains and at the C-terminus giving the protein a net positive charge due to the protonated nitrogen atoms of the amino groups. The charged protein molecules now repel each other and will be more soluble. However, not all precipitations are reversible.

22.39 (a) The number of different tetrapeptides may be calculated by assuming that each of the four amino acids may be located in any of the four positions. The total number is 4 to the power of 4 or 256 possible tetrapeptides.
(b) If all 20 amino acids are available, the total number of possibilities is 20^4 or 160,000.

22.41 Leu, which has a side chain made up of only hydrocarbon, is a nonpolar amino acid. The fewest changes in the protein would result if the Leu were substituted by other nonpolar amino acids like Val or Ile.

22.43 (a) Tropocollagen, the triple helix form of collagen, is considered to be a form of secondary structure.
(b) The collagen fibril form is defined by quaternary structure.
(c) Collagen fibers are considered to be an example of quaternary structure.
(d) The repeating sequence of amino acids refers to the primary structure.

22.45 The carboxylic acid side chain of Glu has a pK_a of 4.25. At low pH values, the side chain carboxyl groups are protonated and thus have no charge. At higher pH (above 4.25), the side chains begin to be deprotonated and become negatively charged. When the polypeptide wraps into an α-helix, this brings the side chains in close proximity (Figure 21.8a). Above a pH of 4.25, the negatively charged side chains interact unfavorably (repel each other) which destabilizes the helix, and the polypeptide forms a random coil. When the side chains are protonated at low pH, the neutral side chains allow for the formation of an α helix.

22.47 Box 1: the carboxyl terminus of the top polypeptide.
Box 2: the amino terminus of the lower polypeptide.
Box 3: a region of antiparallel β-pleated sheet between the top and bottom polypeptides.
Box 4: a section of random coil structure.
Box 5: a region of hydrophobic interactions.
Box 6: an intrachain disulfide bond between Cys residues.
Box 7: a region of α-helix.
Box 8: a salt bridge (ionic interaction) between the side chains of Asp and Lys.
Box 9: a hydrogen bond that is part of the tertiary structure.

22.49 (a) Adult hemoglobin (HbA) has a quaternary structure defined by four subunits, 2 α chains and 2 β chains. (The α and β terms are nomenclature only and do not refer to secondary structure in the protein). Fetal hemoglobin (HbF) also has four subunits, but instead of β subunits, HbF has γ subunits that have a different primary structure

than the β subunits. The actual arrangement of the four subunits is very similar in HbA and HbF.
(b) HbF has a slightly greater affinity for oxygen than does HbA.

22.51 Cytochrome c is an example of a conjugated protein, one that contains, in addition to the protein portion, a non-amino acid portion called a prosthetic group. In cytochrome c the prosthetic group is an iron porphyrin, also called a heme. The two parts are held together by noncovalent interactions. Cytochrome c belongs to the conjugated protein group.

22.53 The urea disrupts hydrogen bonding between protein backbone C=O and H-N groups.

22.55 Cysteine residues in proteins are often linked via intramolecular or intermolecular disulfide bonds, -S-S-. When these bonds are selectively cleaved by reduction, the protein is usually denatured and the cysteine side chains become –SH.

22.57 Ions of heavy metals like silver denature bacterial proteins by reacting with cysteine -SH groups. The proteins, denatured by formation of silver salts, form insoluble precipitates.

22.59 Glutathione is a tripeptide of the amino acids Glu, Cys, and Gly (Chemical Connections 22A). When glutathione is oxidized, two of the molecules react to form a disulfide bond between the two cysteine side chains:

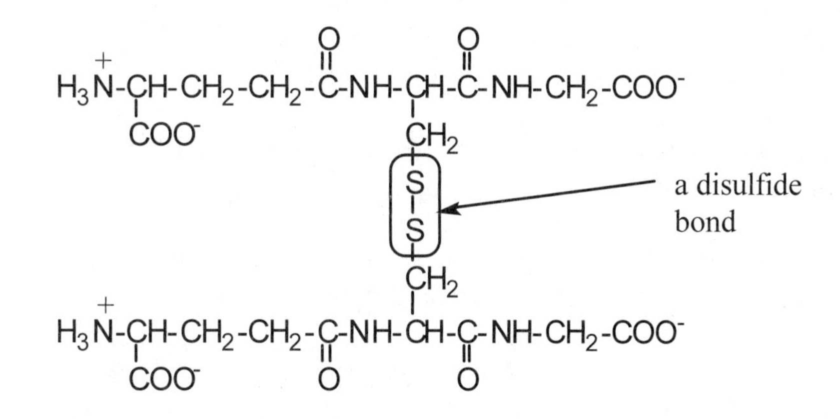

22.61 Hypoglycemic awareness is the sensation of hunger, sweating, and poor coordination that in a diabetic precedes an insulin reaction. This signals a time when insulin levels are too high relative to the blood sugar level (Chemical Connections 22C).

22.63 The α-helical regions of the prion protein are transformed into β-sheets, which tend to aggregate into amyloid plaques. These abnormal prions then stimulate other prions to alter their shape as well, which causes proliferation of the disease.

22.65 The binding curve of myoglobin is hyperbolic. The curve for hemoglobin is sigmoidal. At any partial pressure of oxygen, myoglobin is more highly saturated with oxygen than hemoglobin.

22.67 Diseases associated with amyloid plaques include mad cow disease, Creutzfeld-Jakob disease, and Alzheimer's disease (see Chemical Connections 22E).

22.69 Even if it is feasible, it is not completely correct to call the imaginary process that converts α-keratin to β-keratin denaturation. Any process that changes a protein from α to β requires at least two steps: (1) conversion from α form to a random coil, and (2) conversion from the random coil to the β form. The term denaturation describes only the first half of the process (Step 1). The second step would be called renaturation. So the overall process is called denaturation followed by renaturation. If we assume that the imaginary process actually occurs without passing through a random coil, then the word denaturation does not apply.

22.71 During aging, the triple helices of collagen are crosslinked by the formation of covalent bonds between lysine side chains. This is a form of quaternary structure; the subunits are crosslinked.

22.73 (a) Val and Ile, both with nonpolar side chains, interact by forming hydrophobic bonds.
(b) The side chains of Glu and Lys interact ionically to form a salt bridge.
(c) Hydrogen bonding between hydroxyl groups.
(d) Hydrophobic interactions between alkyl side chains

22.75 Gly has no stereocenter; therefore, it does not exist as enantiomers and does not rotate the plane of polarized light.

22.77 The amino acid Asp has three pK values: 1.88 for the α-carboxyl group; 4.25 for the side-chain carboxyl; 9.6 for the α-amino group. The forms present at pH 2:

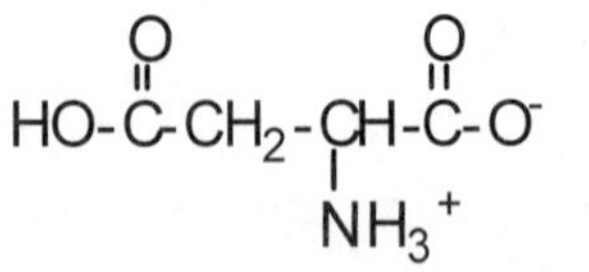

$HO\text{-}C(=O)\text{-}CH_2\text{-}CH(NH_3^+)\text{-}C(=O)\text{-}O^-$

Major form present at pH 2.0; net charge = 0

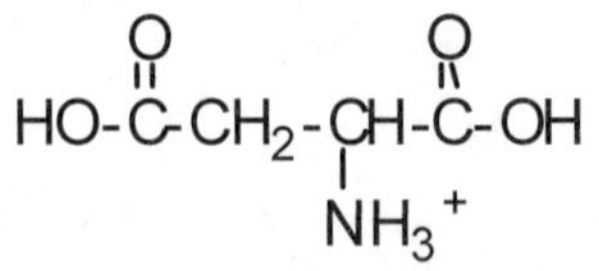

$HO\text{-}C(=O)\text{-}CH_2\text{-}CH(NH_3^+)\text{-}C(=O)\text{-}OH$

Minor form present at pH 2.0; net charge = +1

22.79 The critical amino acids at the active site of an enzyme must be able to catalyze common organic chemistry reactions. Some side chains are more reactive than others. The acidic and basic amino acids are reactive. Serine has a hydroxyl in its side chain that is also very reactive.

22.81 Proteins can be denatured when the temperature is only slightly higher than a particular optimum. For this reason, the health of a warm-blooded animal is dependent on the body temperature. If the temperature is too high, proteins could denature and lose function.

22.83 If you know the genome of an organism, you do not necessarily know the proteins. Not all DNA encodes protein. Some DNA is never transcribed to RNA. Of the DNA that is transcribed to RNA, not all of the RNA is translated to protein. Even if a particular piece of DNA can be used to make RNA and then protein, the protein may not be made all of the time.

23.1 The word catalyst is a general term used to define agents that speed up chemical reactions. Enzymes are natural catalysts, composed of proteins and sometimes RNA, that catalyze reactions in biological cells.

23.3 Lipases, enzymes that catalyze the hydrolysis of ester bonds in triglycerides, are not very specific as they work on many different triglyceride structures with about the same effectiveness. It is predicted that the two triglycerides containing palmitic acid and oleic acid would be hydrolyzed at about the same rate.

23.5 There are thousands of different kinds of biochemicals in an organism and thousands of reactions are required to synthesize and degrade the compounds. The mild conditions inside the cell are not conducive to fast reaction rates so each reaction requires an enzyme for catalysis under physiological conditions.

23.7 Hydrolases use water to break bonds in a substrate usually leading to two smaller products; for example, the enzyme acetylcholinesterase in Table 23.1. Water is always involved in a hydrolase reaction. Lyases catalyze the addition of a group to a double bond or the removal of a group to form a double bond. The added or removed group may be water, but it is not always one of the substrates. An example of a lyase is aconitase in Table 23.1.

23.9 (a) The reactant and product have the same molecular formula, but they have different molecular structures, so they are isomers. The enzyme that catalyzes the reaction is classified as an isomerase.
(b) Water is a reactant that hydrolyzes the two amide bonds in urea. The enzyme is classified as a hydrolase.
(c) Succinate is oxidized and FAD is reduced. The enzyme is classified as an oxidoreductase.
(d) Aspartase catalyzes the addition of a functional group to a double bond. The enzyme is classified as a lyase.

23.11 A cofactor is a nonprotein part of an enzyme that must be present for catalytic activity. Metal ions are often cofactors. When a cofactor is an organic molecule like FAD or heme, it is called a coenzyme.

23.13 Noncompetitive inhibitors slow enzyme activity by binding to sites other than the active site where the substrate (S) binds. A reversible inhibitor (I) binds to and dissociates from the enzyme (E) thus setting up an equilibrium:

$$E + I \rightleftharpoons EI$$

An irreversible inhibitor binds permanently to the enzyme and alters its structure:

$$E + I \longrightarrow E\text{-}I$$

23.15 At very low concentrations of substrate, the rate of an enzyme-catalyzed reaction increases in a linear fashion with increasing substrate concentrations. At higher concentrations of substrate, the active sites of the enzyme molecules become saturated with substrate and the rate of increase is slowed. The rate is at a maximum level when all of the enzyme active sites are occupied with substrate. That maximum rate cannot be exceeded even if more substrate is added.

23.17 (a) Normal body temperature is 37° C (98.6° F) and a fever of 104° F is about 40° C. According to the curve, the bacterial enzyme is more active with the fever.
(b) The activity of the bacterial enzyme decreases if the patient's temperature decreases to 35° C.

23.19 (a) Plot of the pH dependence of pepsin activity:

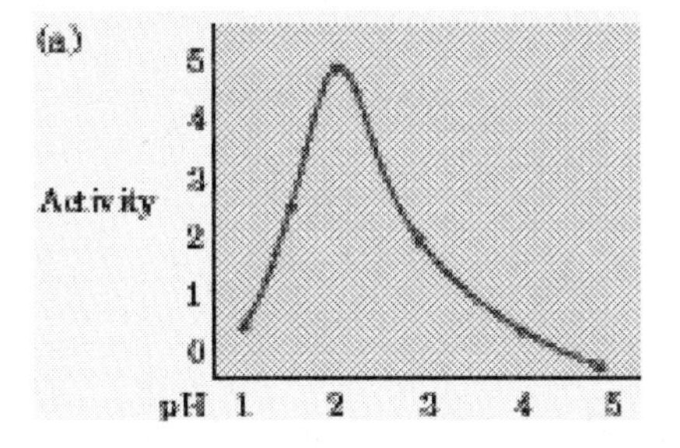

(b) Optimum pH is about 2

(c) Pepsin is predicted to have no catalytic activity at pH 7.4.

23.21 The normal substrate of urease, urea, is a much smaller molecule than diethylurea. Apparently the diethylurea, with its large bulky ethyl groups, is too large to fit into the active site of the enzyme

23.23 The amino acid residues most often found at enzyme active sites are His, Cys, Asp, Arg, and Glu.

23.25 The correct answer is (c). Initially the enzyme does not have just the right shape for strongly binding a substrate, but the shape of the active site changes to better accommodate the substrate molecule.

23.27 Amino acid residues in addition to those at an enzyme active site are present to help form a three-dimensional pocket where the substrate binds. These amino acids act to make the size, shape, and environment (polar or nonpolar) of the active site just right for the substrate.

23.29 At first, one might expect the inhibition of phosphorylase action by caffeine to be a case of traditional noncompetitive inhibition because the inhibitor apparently does not bind at the active site. However, glycogen phosphorylase is known to be an allosteric enzyme and the classical terms competitive and noncompetitive are not used to define inhibition. The term negative modulator is used to define the action of caffeine on the allosteric enzyme, phosphorylase.

23.31 The terms zymogen and proenzyme are both used to define an inactive form of an enzyme.

23.33 The action of a protein kinase is to catalyze the transfer of a phosphate group from ATP to another molecule, in this case a tyrosyl residue of an enzyme:

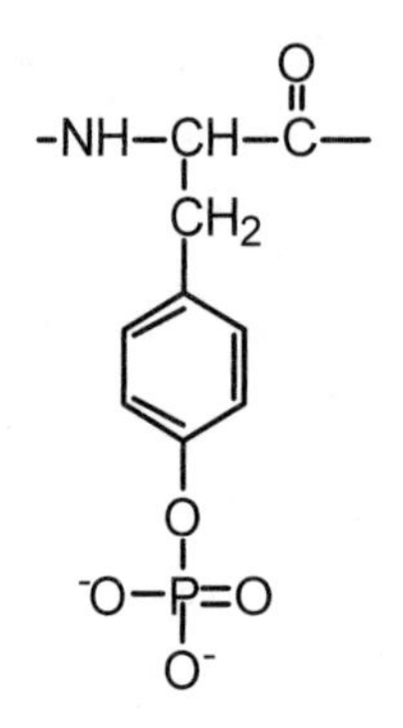

23.35 Glycogen phosphorylase b is converted to glycogen phosphorylase a by the action of an enzyme called phosphorylase kinase. The kinase transfers phosphate groups from ATP to phosphorylase b. Two ATPs are required because the phosphorylase has two subunits, each of which must be phosphorylated:

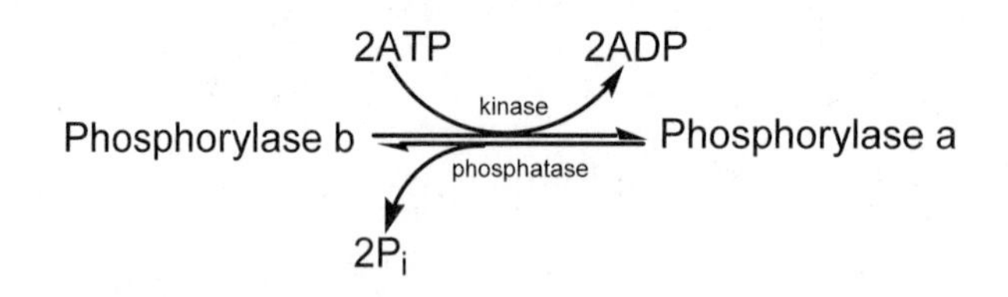

23.37 Glycogen Phosphorylase is controlled by allosteric regulation and by phosphorylation. The allosteric controls are very fast, so that when the level of ATP drops, for example, there is an immediate response to the enzyme allowing more energy to be produced. The covalent modifications by phosphorylation are triggered by hormone responses. They are a bit slower, but more long lasting and ultimately more effective.

23.39 Just like lactate dehydrogenase, there are 5 isozymes of PFK. They are designated as follows: M_4, M_3L, M_2L_2, ML_3, and L_4.

23.41 Two enzymes that increase in serum concentration following a heart attack are creatine phosphokinase and aspartate aminotransferase. Creatine phosphokinase peaks earlier than aspartate aminotransferase, so it would be the best choice in the first 24 hours.

23.43 Serum levels of the enzymes AST and ALT are monitored for diagnosis of hepatitis and heart attack. Serum levels of AST are increased after a heart attack, but ALT levels are normal. In hepatitis, both enzymes are elevated. The diagnosis, until further testing would indicate the patient may have had a heart attack.

23.45 Chemicals present in organic vapors are detoxified in the liver. The enzyme alkaline phosphatase is monitored to diagnose liver problems.

23.47 It is not possible to administer chymotrypsin orally. The stomach would treat it just as it does all dietary proteins: degrade it by hydrolysis to free amino acids. Even if whole, intact molecules of the enzyme were to be present in the stomach, the low pH in the region would not allow activity for the enzyme that has an optimum pH of 7.8.

23.49 A transition state analog is built to mimic the transition state of the reaction. It is not the same shape as the substrate or the product, but rather something in between. The potency with which such analogs can act as inhibitors lends credence to the theory of the induced fit.

23.51 Succinylcholine has a chemical structure similar to that of acetylcholine so they both can bind to the acetylcholine receptor of the muscle end plate. The binding of either choline causes a muscle contraction. However, the enzyme acetylcholinesterase hydrolyzes succinylcholine only very slowly compared to acetylcholine. Muscle contraction does not occur as long as succinylcholine is still present and acts as a relaxant.

23.53 The bacterium that causes stomach ulcers, *H. pylori*, protects itself from stomach acid using the enzyme urease. This enzyme catalyzes the hydrolysis of urea forming ammonia, a weak base. The chemical process continually bathes the bacterial cell with ammonia that neutralizes the stomach acid.

23.55 Antibiotics like amoxillin and tetracycline are used to treat ulcers caused by *H. pylori.*

23.57 In the enzyme pyruvate kinase, the $=CH_2$ of the substrate phosphoenol pyruvate sits in a hydrophobic pocket formed by the amino acids Ala, Gly, and Thr. The methyl group on the side chain of Thr rather than the hydroxyl group is in the pocket. Hydrophobic interactions are at work here to hold the substrate into the active site.

23.59 Researchers were trying to inhibit phosphodiesterases because cGMP acts to relax constricted blood vessels. This was hoped to help treat angina and high blood pressure.

23.61 Phosphorylase exists in a phosphorylated form and an unphosphorylated form, with the former being more active. Phosphorylase is also controlled allosterically by several compounds, including AMP and glucose. While the two act semi-independently, they are related to some degree. The phosphorylated form has a higher tendency to assume the R form, which is more active, and the unphosphorylated form has a higher tendency to be in the less active T form.

23.63 In the processing of cocaine by specific esterase enzymes, the cocaine molecule passes through an intermediate state. A molecule was designed that mimics this transition state, and this transition state analog can be given to a host animal, which then produces antibodies to the analog. When these antibodies are given to a person, they act like an enzyme and degrade cocaine.

23.65 Cocaine blocks the reuptake of the neurotransmitter, dopamine, leading to overstimulation of the nervous system as the neurotransmitter will be around longer.

23.67 (a) Vegetables such as green beans, corn, and tomatoes are heated to kill microorganisms before they are preserved by canning. Milk is preserved by the heating process, pasteurization.
(b) Pickles and sauerkraut are preserved by storage in vinegar (acetic acid).

23.69 The amino acid residues (Lys and Arg) that are cleaved by trypsin have basic side chains, thus are positively charged at physiological pH.

23.71 This enzyme works best at a pH of about 7.

23.73 Chymotrypsin uses water as a substrate to cleave a peptide bond. Thus the reaction is hydrolysis and the enzyme is classified as a hydrolase.

23.75 (a) The reaction involves the oxidation of an alcohol, ethanol, to an aldehyde, acetaldehyde. The enzyme is called ethanol dehydrogenase or in general, alcohol dehydrogenase. It could also be called ethanol oxidoreductase.
(b) The reaction involves the hydrolysis of the ester bond in ethyl acetate. An appropriate name is ethyl acetate esterase or ethyl acetate hydrolase.

23.77 These enzymes, one from brain and one from liver, catalyze the same reaction, but exist in different forms. They would be called isoenzymes or isozymes.

23.79 Caffeine would be more effective for a longer race, like a 10 km race. By stimulating lipases, it would stimulate usage of fatty acids for energy. In a short race, like a one mile race, carbohydrates are the preferred fuel sources. Caffeine actually inhibits the conversion of phosphorylase to the more active R state, so it would be an ineffective aid for an event where carbohydrates are needed for quick energy.

23.81 The structure of RNA makes it more likely to be able to adopt a wider range of tertiary structures, so it can fold up to form globular molecules like protein-based enzymes. It also has an extra oxygen, which gives it an additional reactive group to use in catalysis or an extra electronegative group useful in hydrogen bonding.

24.1 G-protein is a membrane-bound enzyme that catalyzes hydrolysis of GTP to GDP or GMP, whereas GTP is an energy storage molecule.

24.3 A chemical messenger is a biomolecule such as a hormone that binds to a receptor in a cell membrane with the purpose of initiating some change inside the cell. A secondary messenger is a biomolecule like cAMP that transmits the action of the chemical messenger to the inside of the cell, usually with amplification.

24.5 The role of calcium ions in the release of neurotransmitters is most easily understood when one considers a nerve impulse reaching the presynaptic site (Chemical Connections 24A). This causes the opening of ion channels and the entry of calcium ions into the nerve cell that leads to the release of neurotransmitter molecules into the synapse.

24.7 Lactation is controlled by hormones from the anterior pituitary gland. The anterior lobe makes prolactin that stimulates the growth of the mammary gland. The posterior lobe of the pituitary produces the hormone oxytocin, which stimulates milk flow.

24.9 The neurotransmitter acetylcholine is stored in presynaptic vesicles (Chemical Connection 24A). When a nerve impulse arrives, ion channels open and allow calcium ions to enter the nerve cell. The vesicles then fuse with the membrane and release acetylcholine into the synapse where the neurotransmitter molecules bind to receptors on the postsynaptic side. Binding of acetylcholine to the receptors opens ion channels. The flow of ions creates electrical signals.

24.11 Cobratoxin causes paralysis by acting as a nerve system antagonist—it blocks the receptor and interrupts the communication between neuron and muscle cell. The botulin toxin prevents the release of acetylcholine from the presynaptic vesicles.

24.13 Taurine has an amino group and an acidic group hence it may be classified as an amino acid. However, it differs from the amino acids found in proteins. Taurine has a sulfonic acid group instead of a carboxylic acid and the functional groups are on different carbon atoms than in protein amino acids.

24.15 γ-Aminobutyric acid (GABA) is also an amino acid; however, the two functional groups are not bound to the same carbon atom as in the case of protein amino acids. Another name for GABA is 4-aminobutanoic acid.

24.17 (a) Two monoamine neurotransmitters in Table 24.1 are dopamine and serotonin.
(b) They both act by binding to adrenergic receptors, thus continuing the nerve impulse from one neuron to another.
(c) The drug Deprenyl may be used to increase dopamine levels. Serotonin levels are increased by the drug Prozac.

24.19 Protein kinase is activated by interaction with cyclic AMP which dissociates the regulatory and catalytic subunits. The catalytic subunit is then activated by phosphate transfer from ATP (Figure 24.4).

24.21 Product from the MAO-catalyzed oxidation of dopamine:

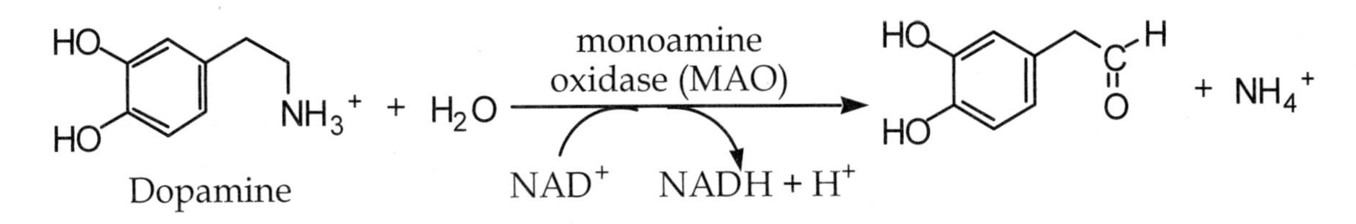

24.23 (a) Amphetamines stimulate adrenergic neurotransmission.
(b) Reserpine decreases the concentration of adrenergic hormones thus causing a sedative effect.

24.25 Removal of the methyl amino group from epinephrine gives a product containing the aldehyde functional group (see Section 24.5F).

24.27 (a) The ion channel is blocked by the ion-translocating protein (see Figure 24.4).
(b) When the blocking protein is phosphorylated, it moves to open the channel.
(c) Cyclic AMP activates the protein kinase that phosphorylates the ion-translocating protein.

24.29 The brain peptides called enkephalins are by chemical nature pentapeptides.

24.31 A kinase catalyzes the phosphorylation of inositol-1,4-diphosphate to inositol-1,4,5-triphosphate:

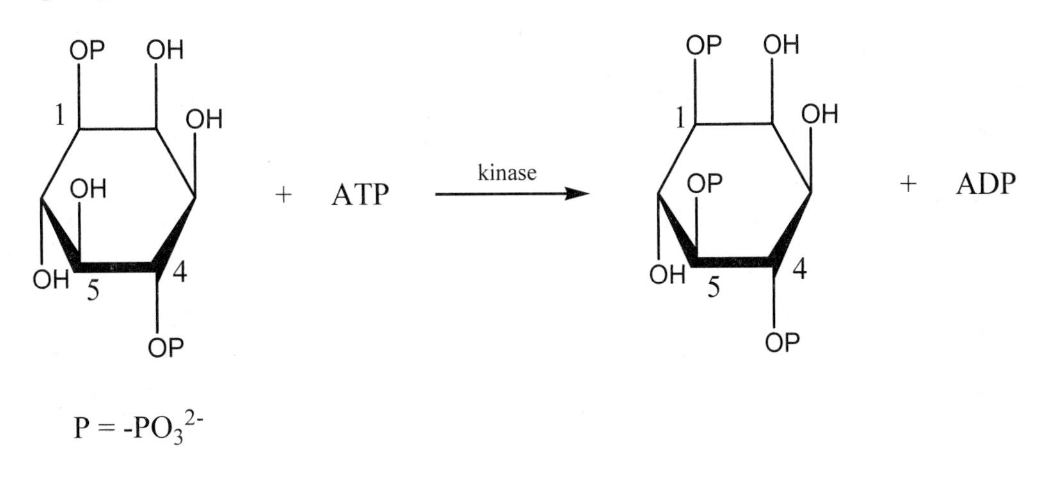

$P = -PO_3^{2-}$

24.33 Most receptors for steroid hormones are located in the cell nucleus. Steroid hormones act by binding to receptors in the nucleus and then affecting the transcription of particular genes.

24.35 In the brain, steroid hormones can act as neurotransmitters. They affect the NMDA and GABA receptors.

24.37 Calmodulin, a calcium ion-binding protein, activates protein kinase II, which catalyzes phosphorylation of other proteins. This process transmits the calcium signal into metabolic activity in the cell.

24.39 (a) Nerve gases cause a variety of effects: sweating, dizziness, vomiting, convulsions, respiratory failure, and often death.
(b) Many nerve gases covalently and irreversibly bind to acetylcholinesterase, thus inactivating the enzyme. With the enzyme dead, the acetylcholine molecules are not degraded and they continue firing the neuron.

24.41 Uncontrolled facial movements (tics) are caused by release of acetylcholine at the nerve-muscle junction of facial muscles. Local injections of botulinum toxin prevent the release of acetylcholine and eliminate the facial movements.

24.43 The plaques are caused by the buildup and deposition of β-amyloid proteins in the brain.

24.45 The neurotransmitter dopamine is deficient in patients of Parkinson's disease. The biochemical cannot be administered orally, because the molecules cannot cross the blood-brain barrier.

24.47 Drugs like Cogentin that block cholinergic receptors are often used to treat the symptoms of Parkinson's disease. These drugs lessen spastic motions and tremors.

24.49 Nitric oxide relaxes the smooth muscle cells that surround blood vessels. This causes increased blood flow in the brain that causes headaches.

24.51 A blocked artery caused by a stroke restricts the blood flow to certain parts of the brain which kills neurons. Adjoining neurons begin to release glutamic acid and NO which kills local cells.

24.53 Insulin-dependent diabetes is caused by insufficient production of insulin by the pancreas. The administration of insulin relieves symptoms of this type of diabetes. Noninsulin-dependent diabetes is caused by a deficiency of insulin receptors or by the presence of inactive receptor molecules. These patients do not respond to injected insulin so they use certain drugs to relieve symptoms.

24.55 The new technique of monitoring glucose levels in the tear relieves the patient of pricking his/her finger many times a day for a blood sample.

24.57 Aldosterone is a steroid hormone that acts by binding to steroid receptors in the nucleus. This complex of steroid and receptor becomes a transcription factor that regulates expression of a gene that codes proteins for mineral metabolism.

24.59 Administering large doses of acetylcholine will ameliorate the overdose of decamethonium bromide. In competitive inhibition the inhibitor can be removed by increase in the substrate concentration. The esterase may regain its activity.

24.61 See structures for α-alanine and β-alanine below. In the normal α-alanine, both the carboxyl and amino groups are bound to the same carbon atom. In β-alanine, the two functional groups are bound to different carbon atoms.

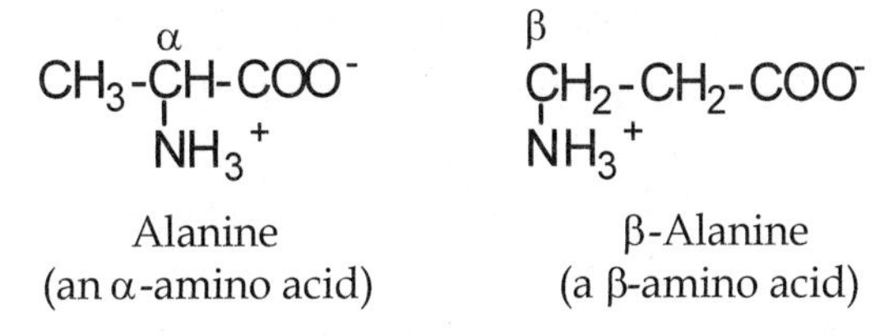

Alanine (an α-amino acid) β-Alanine (a β-amino acid)

24.63 Effects of NO on smooth muscle: dilation of blood vessels and increased blood flow; headaches caused by dilation of blood vessels in brain; increased blood flow in the penis leading to erections.

24.65 Acetylcholineesterase catalyzes the hydrolysis of the neurotransmitter acetylcholine to produce acetate and choline. Acetylcholine transferase catalyzes the synthesis of acetylcholine by bringing together acetylCoA and choline.

24.67 The reaction shown below is the hydrolysis of GTP:

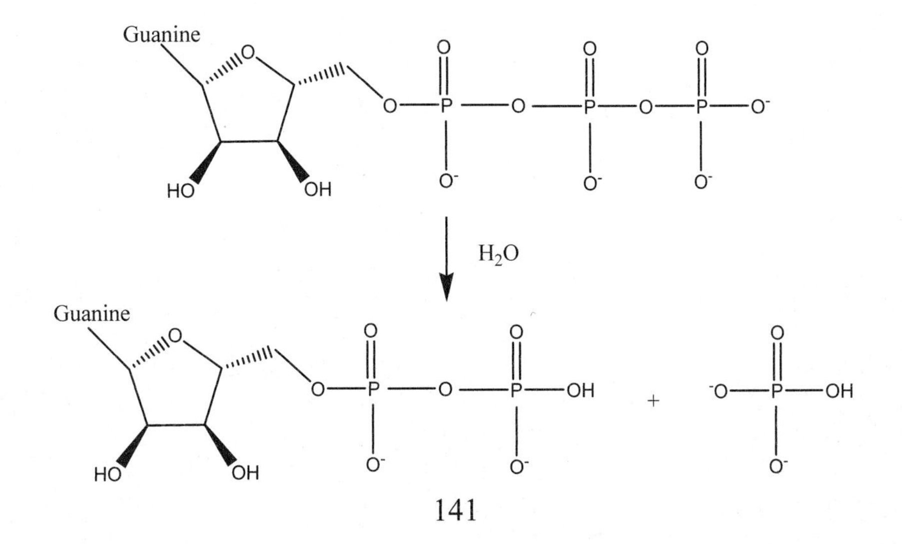

24.69 The drug Ritalin increases serotonin levels. Serotonin has a calming effect on the brain. One advantage of the drug is that, in low doses, it does not enhance the levels of the stimulant, dopamine.

24.71 Proteins are capable of specific interactions at recognition sites, as we saw in chapter 23. This makes proteins ideal for the selectivity receptors must show for particular messengers.

24.73 Adrenergic messengers, such as dopamine, are derivatives of amino acids. For example, a biochemical pathway exists that produces dopamine from the amino acid tyrosine.

24.75 Insulin is a small protein. It would go through protein digestion if taken orally and would not be taken up as the whole protein. In general, peptide hormones are ineffective when administered orally.

24.77 Steroid hormones can directly affect nucleic acid synthesis.

24.79 Chemical messengers vary in their response times. Those that operate over short distances, such as neurotransmitters, have short response times. Their mode of action frequently consists of opening or closing channels in a membrane or binding to a membrane-bound receptor. Hormones must be transmitted in the blood stream, which requires a longer time for them to take effect. Some hormones can and do affect protein synthesis, which makes the response time even longer.

24.81 Having two different enzymes for the synthesis and breakdown of acetylcholine means that the rates of formation and breakdown can be controlled independently.

25.1 Structure of UMP:

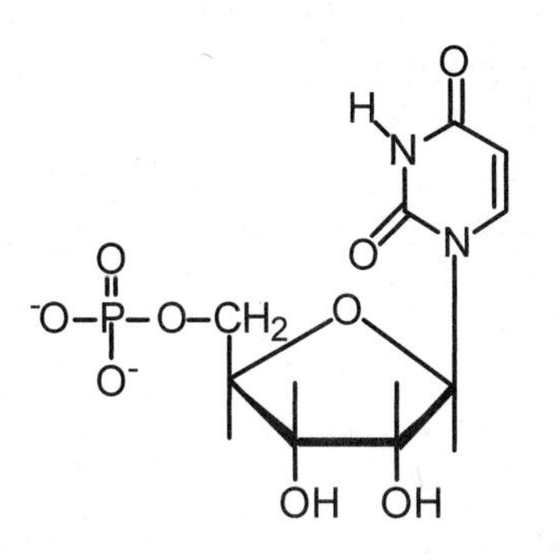

25.3 Scientists and physicians have identified hundreds of hereditary diseases. Sickle cell anemia is one you have studied in detail (Chemical Connections 22D).

25.5 (a) In eukaryotic cells, DNA is located in the cell nucleus and in mitochondria.
(b) RNA is synthesized from DNA in the nucleus, but further use of RNA (protein synthesis) occurs on ribosomes in the cytoplasm.

25.7 DNA has the sugar deoxyribose, while RNA has the sugar ribose. Also, RNA has uracil, while DNA has thymine.

25.9 See Figure 25.1. Thymine and uracil are both based on the pyrimidine ring. However, thymine has a methyl substituent at carbon 5 whereas uracil has a H. All of the other ring substituents are the same.

25.11 (a) Structure of cytidine: (b) Structure of deoxycytidine:

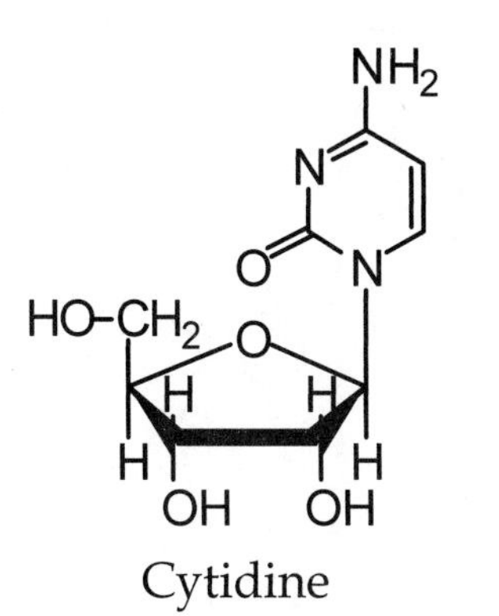

Cytidine

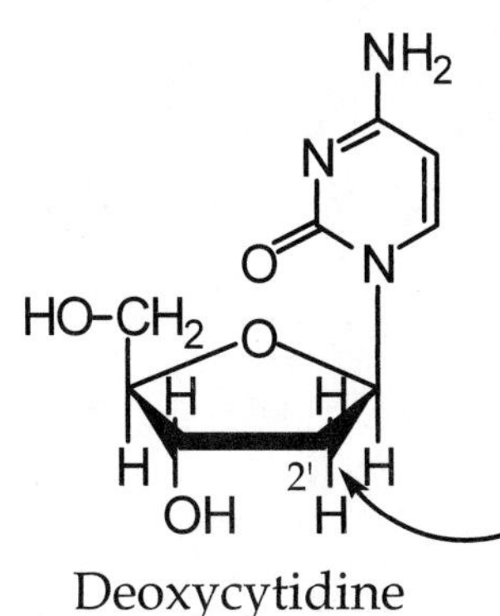

Deoxycytidine

"deoxy" because there is no oxygen at carbon 2'

25.13 D-ribose and 2-deoxy-D-ribose have the same structure except at carbon 2. D-ribose has a hydroxyl group and hydrogen on carbon 2, whereas deoxyribose has two hydrogens (Section 25.2B):

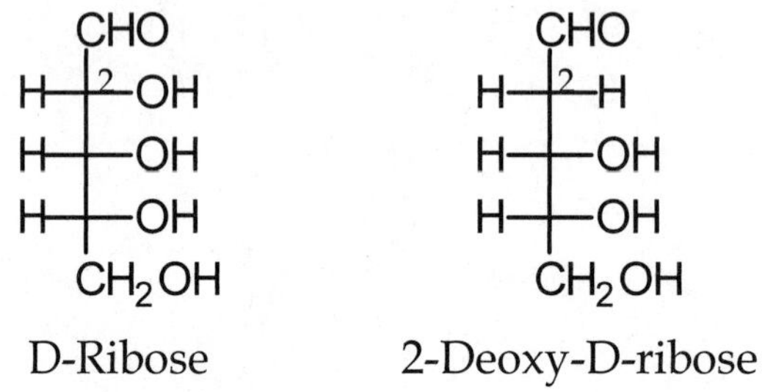

25.15 The name nucleic acid comes from the fact that the nucleosides are linked by phosphate groups, which are the dissociated form of phosphoric acid.

25.17 The bond between the two phosphate groups is called an anhydride bond.

25.19 In RNA, carbons 3' and 5' of the ribose are linked by ester bonds to the phosphates. Carbon 1 is linked to the nitrogen base with the N-glycosidic bond.

25.21 (a) Structure of UDP:

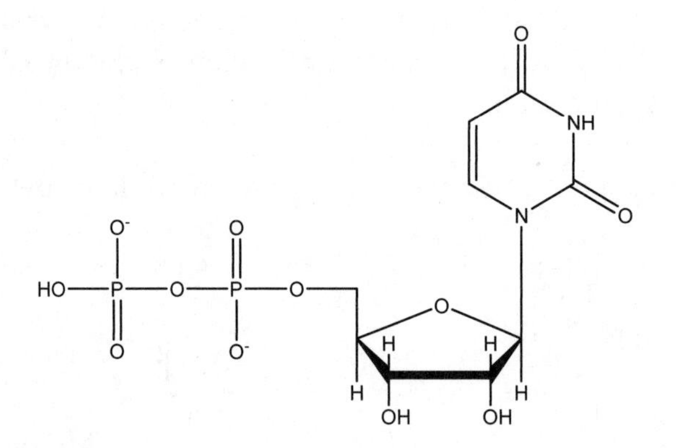

(b) Structure of dAMP:

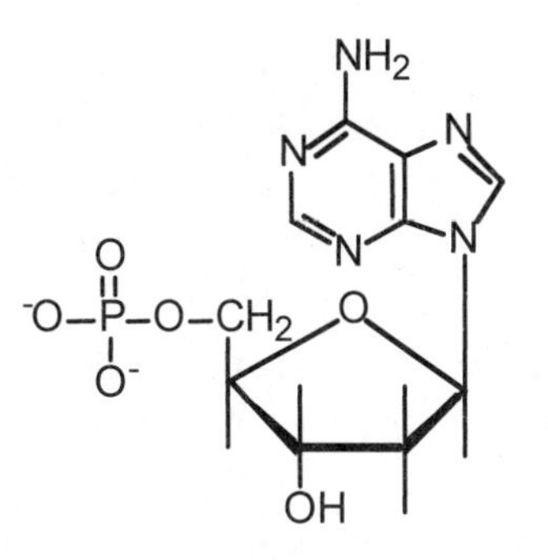

25.23 (a) One end will have a free 5' phosphate or hydroxyl group that is not in phosphodiester linkage. That end is called the 5' end. The other end, the 3' end, will have a 3' free phosphate or hydroxyl group.
(b) By convention the end drawn to the left is the 5' end. A is the 5' end and C is the 3' end.

25.25 Two hydrogen bonds form between uracil and adenine. This base pairing is the same as with adenine and thymine shown in Figure 25.5.

25.27 Histones are proteins with a high content of two basic amino acids, Lys and Arg. Recall that these two amino acid residues in proteins will have positively-charged side chains at physiological pH. On the other hand, DNA at physiological pH will have many negatively charged groups due to the ionized phosphates in the backbone. These two types of molecules will form very strong electrostatic interactions or salt bridges.

25.29 The superstructure of chromosomes is comprised of many elements. DNA and histones combine to form nucleosomes that are wound into chromatin fibers. These fibers are further twisted into loops and minibands to form the chromosome superstructure. (Figure 25.8).

25.31 The double helix is the secondary structure of DNA.

25.33 DNA is wound around histones, collectively forming nucleosomes that are further wound into solenoids, loops, and bands.

25.35 tRNA molecules range in size from 73 to 93 nucleotides per chain. mRNA molecules are larger with an average of 750 nucleotides per chain. rRNA molecules can be as large as 3000 nucleotides.

25.37 All types of RNA will have a sequence complementary to a portion of a DNA molecule because the RNA is transcribed from a gene.

25.39 Ribozymes, or catalytic forms of RNA, are involved in post-transcriptional splicing reactions that cleave larger RNA molecules into smaller more active forms. For example, tRNA molecules are formed in this way. Ribozymes are also part of protein synthesis.

25.41 Small nuclear RNA is part of small nuclear ribonucleoprotein particles which are involved in splicing reactions of other RNA molecules.

25.43 micro RNAs are 22 bases long and prevent transcription of certain genes. Small interfering RNA vary from 22-30 bases and are involved in the degradation of specific mRNA molecules.

25.45 Messenger RNA immediately after transcription contains both introns and exons. The introns are cleaved out by the action of ribozymes that catalyze splicing reactions on the mRNA.

25.47 Satellite DNA, in which short nucleotide sequences are repeated hundreds and even thousands of times, are not expressed into proteins. They are found at the ends and centers of chromosomes and are necessary for stability.

25.49 The greatest single safeguard against errors in DNA duplication is the high level of specificity between the base pairs, A-T and G-C. This protects the replication process from bringing in the wrong bases. This specificity shows strong molecular recognition.

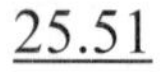

25.51

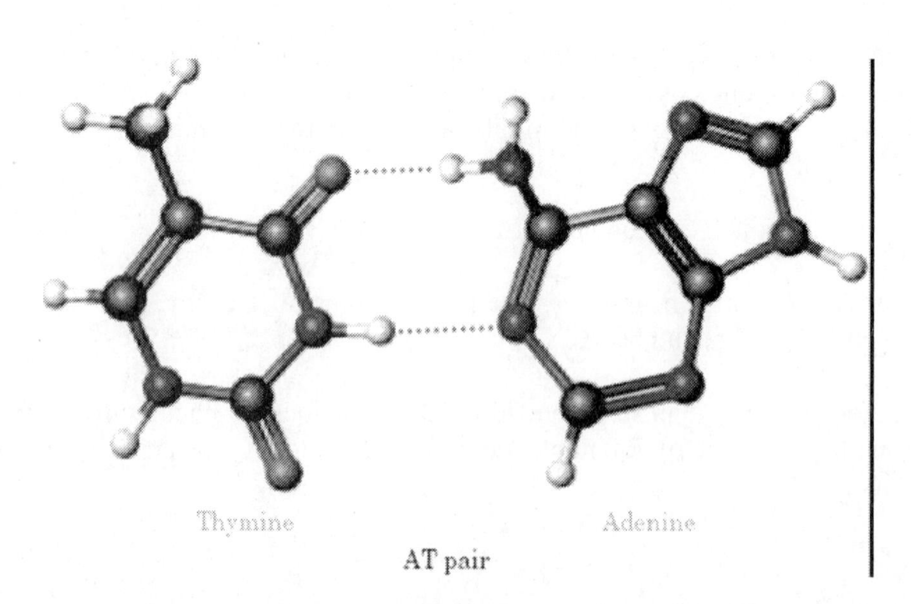

AT pair

25.53 A DNA double helix has four different kinds of bases: A, T, C, and G.

25.55 In semiconservative DNA replication, the original DNA (parental strand) separates and each strand is then used as the template to make its complementary strand. The product is two daughter molecules, each of which have one parental strand and one new strand. This is the mechanism of DNA replication in cells.

25.57 Before the process of DNA duplication can actually begin, the superstructure of the chromosome must be opened so the DNA molecule is accessible for copying. Part of this unwinding process is carried out by histone deacetylase that removes acetyl groups from lysine residue side chains of the histones. The reaction is shown in Section 25.6.

$$\text{Histone}{-}(CH_2)_4{-}NH_3^+ + CH_3{-}COO^- \underset{\text{deacetylation}}{\overset{\text{acetylation}}{\rightleftharpoons}} \text{Histone}{-}(CH_2)_4{-}NH{-}\overset{\overset{\displaystyle O}{\|}}{C}{-}CH_3$$

25.59 Helicases are enzymes that break the hydrogen bonding between the base pairs in double helix DNA and thus help the helix to unwind. This prepares the DNA for the replication process.

25.61 The primers for starting DNA synthesis are short strands of RNA made from ribonucleoside triphosphates. Pyrophosphate (P-P) is a side product formed during the synthesis of the primers.

25.63 The leading strand or continuous strand is synthesized in the 5' to 3' direction so the DNA template is read in the opposite direction.

25.65 The Okazaki fragments are joined together by the enzyme called DNA ligase.

25.67 All nucleotide synthesis, whether it be DNA or RNA, is from the 5' to the 3' direction from the perspective of the chain being synthesized.

25.69 One of the enzymes involved in the DNA base excision repair (BER) pathway is an endonuclease that catalyzes the hydrolytic cleavage of the phosphodiester backbone. The enzyme hydrolyzes on the 5' side of the AP site (Figure 25.15).

25.71 The glycosylase hydrolyzes a β-N-glycosidic bond between the damaged base and the deoxyribose.

25.73 Individuals with the inherited disease XP are lacking an enzyme involved in the NER pathway. They are not able to make repairs in DNA damaged by UV light.

25.75 The 12-nucleotide primer to use is: 5'ATGGCAGTAGGC3'.

25.77 Most anticancer agents work on the method of inhibiting the process of DNA replication. Because cancer cells grow so much faster than normal cells, the cancer cells are impacted much more greatly by inhibiting replication. The anticancer drug fluorouracil causes the inhibition of thymidine synthesis thus disrupting replication.

25.79 DNA polymerase, the enzyme that makes the phosphodiester bonds in DNA, is not able to work at the end of linear DNA. This results in the shortening of the telomeres at each replication. The telomere shortening acts as a timer for the cell to keep track of the number of divisions.

25.81 Because the genome is circular, even if the 5' primers are removed, there will always be DNA upstream that can act as a primer for DNA polymerase to use to synthesize DNA.

25.83 DNA samples from mother, child, and potential father are amplified by PCR, cleaved into fragments using restriction enzymes, and the fragments separated and analyzed by electrophoresis. The figure in CC 25C shows the results of these procedures. Lanes 2, 3, and 4 show the results for the father, child, and mother respectively. The child's DNA contains 6 bands and the mother's has five bands, all of which match those of the child, confirming the mother:child relationship. The band from the alleged father also contains six bands of which only three match the DNA of the child. A child is expected to inherit half of his/her genes from the father. These results do not exclude the potential father because every band in the child's DNA shows up in either the mother lane or the father lane.

25.85 Once a DNA fingerprint is made, each band in the child must come from one of the parents. Therefore, if the child has a band, and the mother does not, then the father must have that band. In this way, possible fathers are eliminated.

25.87 Cytochrome P-450 is a liver enzyme that detoxifies drugs and other synthetic chemicals by adding a hydroxyl group to them. This makes the drugs more water soluble and easier to eliminate in the urine.

25.89 The active site of a ribozyme is a three-dimensional pocket of ribonucleotides where substrate molecules are bound for catalytic reaction. Functional groups for catalysis include the phosphate backbone, ribose hydroxyl groups, and the nitrogen bases.

25.91 (a) The structure of the nitrogen base uracil is shown in Figure 25.1. It is a component of RNA. (b) Uracil with a ribose attached by an N-glycosidic bond is called uridine.

25.93 Native DNA is the largest nucleic acid.

25.95 Mol % A = 29.3; Mol % T = 29.3; Mol % G = 20.7; Mol % C = 20.7.

25.97 RNA synthesis, like DNA synthesis, is 5' to 3' from the perspective of the growing chain. This is because the nature of the polymerase reaction is the nucleophilic attack of the 3' hydroxyl group on the 5' phosphate of another nucleotide.

25.99 DNA replication requires a primer, which is RNA. Since RNA synthesis does not require a primer, it makes sense that RNA must have preceded DNA as a genetic material. This added to the fact that RNA has been shown to be able to catalyze reactions means that RNA can be both an enzyme and a heredity molecule.

26.1 Transcription begins when the DNA double helix begins to unwind at a point near the gene that is to be transcribed. The superstructures of DNA (chromosomes, chromatin, etc.) break down to the nucleosome (histones plus DNA). Binding proteins interact with the nucleosomes making the DNA less dense and more accessible. The helicase enzymes then begin to unwind the double helix.

26.3 Valine + tRNA (specific for Val) + ATP

26.5 Answer (c); gene expression refers to both processes, transcription and translation.

26.7 Protein translation occurs on the ribosomes.

26.9 Helicases are enzymes that catalyze the unwinding of the DNA double helix prior to transcription. The helicases break the hydrogen bonds between base pairs.

26.11 The termination signal for transcription is at the 5' end of the DNA.

26.13 The "guanine cap" methyl group is located on nitrogen number 7 of guanine.

26.15 A codon, the three-nucleotide sequence that specifies amino acids for protein synthesis, is located on a mRNA molecule.

26.17 The main subunits are the 60S and the 40S ribosomal subunits, although these can be dissociated into even smaller subunits.

26.19 Each triplet sequence of the DNA has the code for a single amino acid in the protein. Therefore, the total number of amino acids is $981 \div 3 = 327$.

26.21 Leucine, arginine, and serine have the most, with 6 codons apiece. Methionine and tryptophan have the fewest with one apiece.

26.23 The amino acid for protein translation is linked via an ester bond to the 3' end of the tRNA. The energy for producing the ester bond comes from breaking two, energy-rich phosphate anhydride bonds in ATP (producing AMP and two phosphates).

26.25 (a) The 40S subunit in eukaryotes forms the pre-initiation complex with the mRNA and the Met-tRNA that will become the first amino acid in the protein.
(b) The 60S subunit binds to the pre-initiation complex and brings in the next aminoacyl-tRNA. The 60S subunit contains the peptidyl transferase enzyme.

26.27 Elongation factors are proteins that participate in the process of tRNA binding and movement of the ribosome on the mRNA during the elongation process in translation.

26.29 There is a special tRNA molecule used for initiating protein synthesis. In Prokaryotes, this is $tRNA^{fmet}$, which will carry a formyl-methionine. In Eukaryotes, there is a similar one, but it carries methionine. However, this tRNA carrying methionine for the initiation of synthesis is different from the tRNA carrying methionine for internal positions.

26.31 This is because it has been shown that there are no amino acids in the vicinity of the nucleophilic attack that leads to peptide bond formation. Therefore, the ribosome must be using its RNA portion to catalyze the reaction, so it is a type of enzyme called a ribozyme.

26.33 The parts of DNA involved in control of transcription are promoters, enhancers, silencers, and response elements. Molecules that bind to the DNA are RNA polymerase and a variety of transcription factors.

26.35 The active site of aminoacyl-tRNA synthases (AARS) contains the sieving portions that act to make sure that each amino acid is linked to its correct tRNA. There are two sieving steps that work on the basis of the size of the amino acid.

26.37 Both are DNA sequences that bind to transcription factors. The difference is largely due to our own understanding of the big picture. A response element is understood to control a set of responses in a particular metabolic context, such as a response element that activates several genes when the organism is challenged metabolically by heavy metals, by heat, or by a reduction in oxygen pressure.

26.39 Proteosomes are cylindrical assemblies of a number of protein subunits with proteolytic activity. Proteosomes play a role in post-translational degradation of damaged proteins. Proteins damaged by age or proteins that have misfolded are degraded by the proteosomes.

26.41 (a) Silent mutation: assume the DNA sequence is TAT on the coding strand, which will lead to UAU on the mRNA. Tyrosine is incorporated into the protein. Now assume a mutation in the DNA to TAC. This will lead to UAC in mRNA. Again, the amino acid will be tyrosine.
(b) Lethal mutation: the original DNA sequence on the coding strand is GAA, which will lead to GAA on mRNA. This codes for the amino acid glutamic acid. The DNA mutation TAA will lead to UAA, a stop signal that incorporates no amino acid. This could mean a vital protein is not made and could lead to disease or even a non-viable organism.

26.43 Yes, a harmful mutation may be carried as a recessive gene from generation to generation with no individual demonstrating symptoms of the disease. Only when both parents carry recessive genes does an offspring have a 25% chance of inheriting the disease.

26.45 Restriction endonucleases are enzymes that recognize specific sequences on DNA and catalyze the hydrolysis of phosphodiester bonds in that region thus cleaving both strands of the DNA (see Section 26.8). These enzymes are useful for the preparation of recombinant DNA.

26.47 Mutation by natural selection is an exceedingly long, slow process that has occurred for centuries. Each natural change in the gene has been ecologically tested and found usually to have a positive effect or the organism is not viable. Genetic engineering, where a DNA mutation is done very fast, does not provide sufficient time to observe all of the possible biological and ecological consequences of the change.

26.49 Vitravene, an antisense drug used by AIDS patients, is made up of nucleotides, but it has a backbone different from the normal phosphodiester linkages. It has a sulfur atom substituted for an oxygen atom in each phosphate group. This functional group change slows the physiological degradation of the drug.

26.51 The viral coat is a protective protein covering around a virus particle.

26.53 CREB binds to the CRE (cyclic AMP response element). It has a phosphorylated form and an unphosphorylated form. When it is phosphorylated, it can bind the CREB binding protein (CBP), which links the CREB to the RNA Polymerase. This form is what stimulates transcription of those genes controlled by CRE.

26.55 An invariant site is a location in a protein that has the same amino acid in all species that have been studied. Studies of invariant sites help establish genetic links and evolutionary relationships.

26.57 The protein p53 is a tumor suppressor. When the protein's gene is mutated, the protein no longer controls replication and the cell begins to grow at an increased rate.

26.59 (a) Transcription: the units include the DNA being transcribed, the RNA polymerases, and a variety of general transcription factors.
(b) Translation: mRNA, ribosomal subunits, aminoacyl-tRNA, initiation factors, and elongation factors.

26.61 Hereditary diseases cannot be prevented, but genetic counseling can help people understand the risks involved in passing a mutated gene to offspring.

26.63 (a) Plasmid: a small, closed circular piece of DNA found in bacteria. It is replicated in a process independent of the bacterial chromosome.
(b) Gene: a section of chromosomal DNA that codes for a particular protein molecule or RNA.

26.65 Each of the amino acids has four codons. All the codons start with G. The second base is different for each amino acid. The third base may be any of the four possible bases. The distinguishing feature for each amino acid is the second base.

26.67 The hexapeptide is Ala-Glu-Val-Glu-Val-Trp.

27.1 Chemical energy present in food molecules is extracted and converted to a usable form by the process of catabolism. The energy derived from the degradation of different types of molecules is collected in the form of the energy-rich molecule, ATP.

27.3 (a) Mitochondria have two membranes, a highly-folded inner membrane, and an outer membrane (see Figure 27.3).
(b) The outer membrane is permeable to the diffusion of small ions and molecules. Special transport processes are required to move molecules through the inner membrane.

27.5 Cristae are the folds that are present in the inner mitochondrial membrane. The folds provide extensive surface area for the concentration of enzymes and other components required for metabolism.

27.7 Each ATP molecule has two phosphate anhydride bonds that release a substantial amount of energy when they are hydrolyzed during metabolic processes (Figure 27.5):

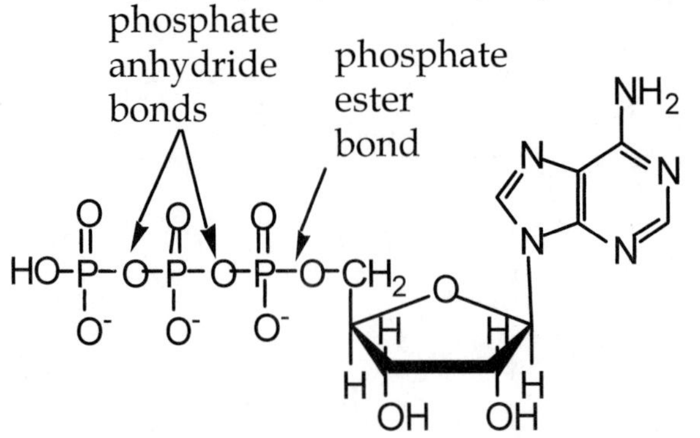

27.9 In reactions (a) and (b) the same type of anhydride bond is hydrolyzed. When the reactions are measured under standard conditions, the energy yield is about the same, 7.3 kcal/mole. In the cell, however, the concentration of ADP is very low, compared to ATP, and ADP is rarely used for energy.

27.11 The chemical bond between the ribitol and flavin in FAD is an amine (see Figure 27.6).

27.13 Two nitrogen atoms in the flavin ring that are linked to carbon (N=C) are reduced to yield $FADH_2$.

27.15 (a) The most important carrier of phosphate groups is ATP.
(b) The most important carriers of hydrogen ions and electrons from redox reactions are NADH and $FADH_2$.
(c) Coenzyme A carries acetyl groups.

27.17 An amide linkage is formed to bring together the amine group of mercaptoethylamine and the carboxyl group of pantothenic acid (Figure 27.7).

27.19 No, the reactive part of CoA is the thiol group (-SH) at the end of the molecule.

27.21 Most fats and carbohydrates are degraded in catabolism to the compound acetyl CoA.

27.23 α-Ketoglutarate is the only C-5 intermediate in the citric acid cycle.

27.25 FAD is used in the citric acid cycle as a coenzyme to oxidize succinate to fumarate.

27.27 Fumarase is classified as a lyase because it catalyzes the addition of water to a double bond.

27.29 ATP is not produced directly by the citric acid cycle. GTP, a reactive molecule with the same amount of energy as ATP, is produced in Step 5 and may be used for some energy-requiring processes.

27.31 The stepwise degradation and oxidation of acetyl CoA in the citric acid cycle is very efficient in the extraction and collection of energy. Rather than occurring in one single burst, the energy is released in small increments and collected in the form of reduced cofactors, NADH and $FADH_2$.

27.33 Carbon-carbon double bonds are present in the citric acid cycle intermediates fumarate and *cis*-aconitate.

27.35 When α-ketoglutarate is oxidized in Step 4 of the citric acid cycle, the electrons are transferred to NAD^+ to make the reduced form, NADH.

27.37 The mobile carriers of electrons in the electron transport chain are coenzyme Q and cytochrome *c*.

27.39 The proton translocator ATPase is a complex that resembles a rotor engine. It rotates every time a proton passes through the inner membrane (Figure 27.10).

27.41 During oxidative phosphorylation water is formed from protons, electrons, and oxygen on the matrix side of the inner membrane. This occurs when electrons are shuttled through complex IV (Figure 27.10).

27.43 (a) For each pair of protons translocated through the ATPase complex, one molecule of ATP is generated. Each pair of electrons that enters oxidative phosphorylation at complex I yields three ATP.

(b) Each C-2 fragment, which represents the carbons in acetyl CoA, yields 12 ATP.

27.45 Protons are translocated through a proton channel formed by the Fo part of the ATPase which has 12 protein subunits embedded in the inner membrane (Figure 26.10).

27.47 The catalytic unit of ATPase is composed of α and β subunits (Figure 27.10). This part of the ATPase catalyzes the formation of ATP:

$$ADP + P_i \longrightarrow ATP + H_2O$$

27.49 The energy, in kcal, from the oxidation of 1 g of acetate by the citric acid cycle:
Molecular wt. acetate = 59 g/mole. 1 g of acetate = 1 ÷ 59 = 0.017 mole. Each mole of acetate produces 12 moles of ATP (See Problem 27.43(b)).
0.017 mole x 12 = 0.204 mole of ATP.
0.204 mole of ATP x 7.3 kcal/mole = 1.5 kcal.

27.51 (a) Muscle contraction takes place when thick protein filaments called myosin slide past thin protein filaments called actin. The hydrolysis of ATP by myosin, an ATPase, drives the alternating association and dissociation of actin and myosin. This causes the contraction and relaxation of muscles (Section 27.8C).
(b) The energy in muscle contraction comes from ATP.

27.53 Glycogen phosphorylase is activated by phosphorylation of serine residues in the protein subunits. The phosphoryl groups are transferred from ATP (see Chemical Connection 23E). Phosphorylase is also activated by allosteric effectors.

27.55 The antibiotic oligomycin inhibits the catalytic subunits of ATPase, thus it stops phosphorylation of ADP. Although it acts as an effective antibiotic, its toxic effect on ATPase does not allow its use in humans.

27.57 (a) Cytochrome P-450 is an enzyme that catalyzes the hydroxylation of various natural and synthetic substrates. The hydroxyl group is derived from molecular oxygen.
(b) The source of the reactant oxygen is the air we breathe. Most (90%) of the oxygen we breathe is used for respiration.

27.59 Number of g of acetic acid that must be metabolized to yield 87.6 kcal of energy:
87.6 kcal ÷ 7.3 kcal/mole ATP = 12 moles of ATP. 12 moles of ATP are released from the oxidation of 1 mole of acetate. 1 mole of acetate (MW = 60) = 60 g of acetic acid.

27.61 The mechanical energy generated from the translocation of protons in oxidative phosphorylation is first displayed in the rotating ion channel of ATPase.

27.63 Citrate and malate, intermediates in the citric acid cycle, both have carboxyl groups and a hydroxyl group.

27.65 Myosin, the thick filament in muscle, is an enzyme that acts as an ATPase.

27.67 Isocitrate has two stereocenters.

27.69 The proton channel is located in the F_o unit of ATPase that is embedded in the inner membrane.

27.71 All the sources of energy used for ATP synthesis are not completely known at this time. Most of the energy comes from proton translocation. Some energy for proton pumping comes also from breaking the covalent bond of oxygen (reduction of oxygen to water).

27.73 Succinate dehydrogenase catalyzes the oxidation of succinate to fumarate. FAD becomes reduced in the reaction and then transfers its electrons directly to the electron transport chain at complex II.

27.75 Citrate isomerizes to isocitrate to convert a tertiary alcohol to a secondary alcohol. Tertiary alcohols cannot be oxidized, but secondary alcohols can be oxidized to produce a keto group, which is necessary to continue the pathway.

27.77 Iron is found in iron-sulfur clusters in proteins and is also part of the heme group of cytochromes.

27.79 Mobile electron carriers transfer electrons on their path from one large, less mobile, protein complex to another.

27.81 ATP and reducing agents such as NADH and $FADH_2$, generated by the citric acid cycle, are needed for biosynthetic pathways. Also, many of the intermediates of the citric acid cycle are drawn off as part of biosynthetic pathways.

27.83 Biosynthetic pathways are likely to be ones of reduction because their net effect is to reverse catabolism, which is oxidative.

27.85 ATP is not stored in the body. It is hydrolyzed to provide energy for many different kinds of processes and thus turns over rapidly. While the body produces kilograms of ATP every day, it also uses kilograms every day. The average life span of an individual molecule of ATP is less than a second.

27.87 The citric acid cycle generates NADH and $FADH_2$, which are linked to oxygen by the electron transport chain.

<u>28.1</u> According to Table 28.2 the ATP yield from stearic acid is 146 ATP. This makes 146/18 = 8.1 ATP/carbon atom.

For lauric acid (C_{12}):

Step *1* Activation	-2 ATP
Step *2* Dehydrogenation five times	10 ATP
Step *3* Dehydrogenation five times	15 ATP
Six C_2 fragments in common pathway	<u>72 ATP</u>
Total	95 ATP

95/12 = 7.9 ATP per carbon atom for lauric acid. Thus stearic acid yields more ATP/C atom.

<u>28.3</u> The major use of amino acids is in the synthesis of proteins. Proteins from ingested food are hydrolyzed and the amino acids are used to rebuild proteins that the body constantly degrades. We cannot store amino acids so we need a constant supply in our diet.

<u>28.5</u> The step referred to is # 4, the aldolase-catalyzed cleavage of fructose 1,6-bisphosphate to glyceraldehyde 3-phosphate and dihydroxyacetone phosphate. Glyceraldehyde 3-phosphate metabolism continues immediately in glycolysis (Step # 5), but the dihydroxyacetone phosphate must first be isomerized to glyceraldehyde 3-phosphate by an isomerase. Glyceraldehyde 3-phosphate and dihydroxyacetone phosphate are in equilibrium and removal of glyceraldehyde 3-phosphate by glycolysis drives the isomerization reaction.

<u>28.7</u> (a) The steps in glycolysis that need ATP are # 1, phosphorylation of glucose and # 3, the phosphorylation of fructose 6-phosphate.
(b) The steps that yield ATP are # 6, catalyzed by phosphoglycerate kinase and # 9, catalyzed by pyruvate kinase.

<u>28.9</u> ATP is a negative modulator for the allosteric, regulatory enzyme phosphofructokinase, Step # 3, as well as for the enzyme pyruvate kinase, step #9.

<u>28.11</u> The oxidation of glucose 6-phosphate by the pentose phosphate pathway produces NADPH. This reduced cofactor is necessary for many biosynthetic pathways, but especially for the synthesis of essential fatty acids, and as a defense against oxidative damage.

<u>28.13</u> The anaerobic degradation of a mole of glucose leads to two moles of lactate. Therefore, three moles of glucose produce six moles of lactate.

28.15 Using the data in Table 28.1, a net yield of 2 ATPs are directly produced by the glycolysis of glucose (glucose to pyruvate). There is the initial expenditure of two ATPs in the first three steps of glycolysis. Then steps 6 and 9 produce 4 ATPs, for a net yield of 2. Most of the ATP from glucose degradation comes from oxidation of the reduced cofactors, NADH and $FADH_2$, linked to respiration and the common pathway.

28.17 (a) Fructose catabolism by glycolysis in the liver yields two ATPs just like glucose. (b) Glycolytic breakdown of fructose in muscle also yields two ATPs per fructose.

28.19 Enzymes that catalyze the phosphorylation of substrates using ATP are called kinases. Therefore, the enzyme that transforms glycerol to glycerol 1-phosphate is called glycerol kinase.

28.21 (a) The enzymes are thiokinase and thiolase. (b) "Thio" refers to the presence of the element sulfur. (c) Both of these enzymes use Coenzyme A that contains a reactive thiol group, –SH, as a substrate.

28.23 Each turn of fatty acid β-oxidation yields one C-2 fragment (acetyl CoA), one $FADH_2$, and one NADH. Therefore, the total yield from three turns is three acetyl CoA, three $FADH_2$, and three NADH. There is still a six-carbon portion of lauric acid left, hexanoyl CoA.

28.25 Using the data from Table 28.2, the yield from the oxidation of one mole of myristic acid is 112 moles of ATP. The process requires six turns of β oxidation and produces 7 moles of acetyl CoA.

28.27 Under normal conditions, the body preferentially uses glucose as an energy source. When a person is well fed (balance of carbohydrates and fats and proteins), fatty acid oxidation is slowed and the acids are linked to glycerol and are stored in fat cells for use in times of special need. Fatty acid oxidation becomes important when glucose supplies begin to be depleted, for example, during extensive physical exercise or fasting or starvation.

28.29 The transformation of acetoacetate to β-hydroxybutyrate is a redox reaction using the cofactor, NADH. Acetone is produced by the spontaneous decarboxylation of acetoacetate.

28.31 Oxaloacetate produced from the carboxylation of PEP normally enters the citric acid cycle at Step 1. As we will learn in the next chapter, oxaloacetate may also be used to synthesize glucose.

28.33 Oxidative deamination of alanine:

$$CH_3-\underset{\underset{NH_3^+}{|}}{CH}-COO^- + NAD^+ + H_2O \longrightarrow CH_3-\underset{\underset{O}{\|}}{C}-COO^- + NADH + H^+ + NH_4^+$$

Alanine Pyruvate

28.35 One of the nitrogen atoms in urea comes originally from an ammonium ion through the intermediate carbamoyl phosphate (steps 1 and 2 in the urea cycle). The ammonium ion was probably released from an amino acid by oxidative deamination. The other nitrogen atom of urea comes from aspartate that enters the urea cycle at step 3.

28.37 (a) The toxic product from the oxidative deamination of Glu is the ammonium ion.
(b) The ammonium ion is converted to urea by the urea cycle and eliminated in the urine.

28.39 Tyrosine is considered a glucogenic amino acid because pyruvate can be converted to glucose when the body needs it. Any amino acid with an easy pathway to pyruvate will be considered glucogenic.

28.41 During initial hemoglobin catabolism, the heme group and globin proteins are separated. The globins are hydrolyzed to free amino acids that are recycled and the iron is removed from the porphyrin ring and saved in the iron-storage protein, ferritin, for later use.

28.43 During exercise, normal glucose catabolism shifts to a greater production of lactate rather than conversion of pyruvate to acetyl CoA and entry into the citric acid cycle. This shift in metabolism is a result of a depletion of oxygen supplies. A build-up of lactate in muscle leads to a lowering of pH which effects myosin and actin action.

28.45 The acidic nature of the ketone bodies lowers blood pH. This increase in proton concentration is neutralized by the bicarbonate/carbonic acid buffer system present in blood (Section 9.11D and Chemical Connections 9D).

28.47 It is necessary to tag proteins for destruction so that the ones that are no longer needed can be turned over without degrading proteins that are needed.

28.49 Ubiquitin is linked to targeted proteins by forming an amide bond between the carboxyl terminus of ubiquitin (Gly) to a side-chain amino group on a lysine residue of the doomed protein.

28.51 When phenylalanine accumulates, it is converted to phenylpyruvate via transamination:

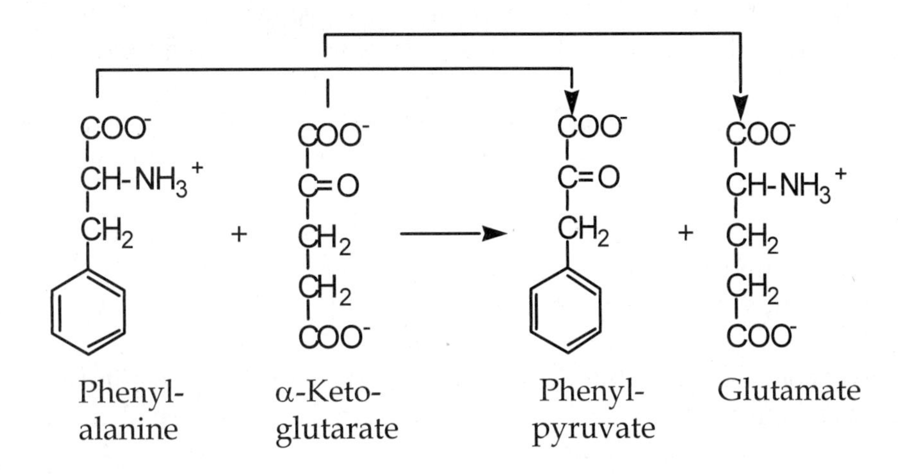

28.53 The presence of a high concentration of ketone bodies in the urine of a patient is usually indicative of diabetes. However, before the disease can be confirmed, other, more detailed tests must be completed as fasting or starvation or special dieting can also increase ketone bodies.

28.55 (a) NAD^+ participates in step # 5, the oxidation of glyceraldehyde 3-phosphate to 1,3-bisphosphoglycerate and step # 12, the oxidation of pyruvate to acetyl CoA.
(b) NADH participates in steps # 10 and # 11, reduction of pyruvate to ethanol and lactate, respectively.
(c) If one considers the path from glucose to lactate, then there is no net gain of the cofactors. If one considers the path from glucose to pyruvate, then there is a gain of two NADH per glucose. Pyruvate to acetyl CoA would add another two NADH per glucose.

28.57 The amino acids Ala, Gly, and Ser are glucogenic; that is, their carbon atoms may be used to synthesize glucose, thus relieving hypoglycemia (low blood sugar).

28.59 Products of the transamination reaction of Ala and oxaloacetate:

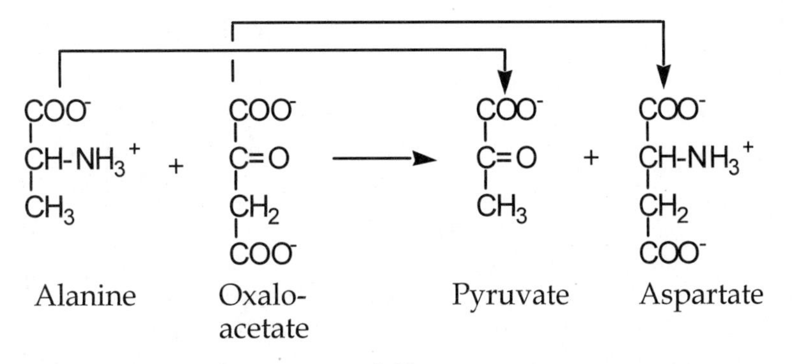

28.61 A number of metabolic processes could occur with the radioactive fatty acid so different molecules should be analyzed. Some of the radioactive fatty acid could be stored in triglycerides in fat tissue; some radioactivity would be in acetyl CoA after β oxidation of the fatty acids; and some would be in carbon dioxide (released from citric acid cycle).

28.63 The urea cycle is an energy-consuming pathway as it requires 3 molecules of ATP (four phosphate anhydride bonds) to produce a single urea from carbon dioxide and two ammonium ions.

28.65 (a) The β-oxidation of lauric acid (12 carbons) requires 5 turns.
(b) Palmitic acid (16 carbons) requires 7 turns.

28.67 The conversion of pyruvate the acetyl CoA is part of aerobic metabolism. The other two possibilities, conversion to lactate or (in some organisms) to ethanol are anaerobic.

28.69 When people are on severely restricted diets, they use up their carbohydrate stores quickly. When a person is metabolizing fats and not carbohydrates, ketone bodies will be produced.

28.71 The nitrogen portion of amino acids tends to be excreted, but carbon skeletons are degraded to yield energy or, alternatively, to serve as building blocks for biosynthetic processes.

28.73 Energy is required to form amide bonds. This process is the opposite of the first step in the digestion of proteins.

28.75 All catabolic pathways produce compounds that eventually enter the citric acid cycle. Many intermediates of the citric acid cycle are starting points for biosynthetic pathways.

28.77 Citric acid is an excellent nutrient, since it can be completely degraded to carbon dioxide and water. It enters the mitochondrion easily.

29.1 Your text states in Section 29.1 several reasons why anabolic pathways are different from catabolic: (1) Duplication of pathways adds flexibility. If the normal biosynthetic pathway is blocked, the body can use the reverse of the catabolic pathway to make the necessary metabolites. (2) Separate pathways allow the body to overcome the control of reactions by reactant concentration (Le Chatelier's principle). (3) Different pathways provide for separate regulation of each pathway. Although there are many differences between anabolism and catabolism, we will also note similarities that allow for coordinated regulation and proper balancing of concentrations.

29.3 The cellular concentration of inorganic phosphate, a reagent used in phosphorylation reactions, is very high; therefore, the reaction is driven in the direction of glycogen breakdown. In order to ensure the presence of glycogen when needed, it must have an alternate synthetic pathway.

29.5 The major difference between the overall reactions of photosynthesis and respiration is the direction of the reactions. They are the reverse of each other:

$$6CO_2 + 6H_2O \rightarrow C_6H_{12}O_6 + 6O_2 \quad \text{photosynthesis}$$
$$C_6H_{12}O_6 + 6O_2 \rightarrow 6CO_2 + 6H_2O \quad \text{respiration}$$

29.7 A compound that can be used for gluconeogenesis:
(a) From glycolysis: pyruvate
(b) From the citric acid cycle: oxaloacetate
(c) From amino acid oxidation: alanine

29.9 The brain obtains most of its energy from glucose that is supplied by the blood. The brain has little or no capacity to store glucose in glycogen. During starvation, glucose for the brain will come from glycogen that is stored primarily in the liver. Since glucose concentrations are very low in starvation, the liver glycogen is synthesized from glucose that is produced from pyruvate, lactate, amino acids, etc. The glucose is formed by gluconeogenesis. Brain cells are also able to obtain some energy from degradation of the ketone bodies.

29.11 Maltose is a disaccharide that is composed of two glucose units linked by an α-1,4-glycosidic bond (Section 20.4C). We know that in glycogen synthesis, the UDP-glucose can combine with another glucose to add to the glycogen chain. Therefore, we could envision a similar reaction to make maltose. The enzyme might be called maltose synthase:

$$\text{UDP-glucose} + \text{glucose} \longrightarrow \text{maltose} + \text{UDP}$$

29.13 Uridine triphosphate (UTP) is a nucleoside triphosphate similar to ATP. The constituents are: a nitrogen base, uracil; a sugar, ribose; and three phosphates.

29.15 (a) The biosynthesis of fatty acids occurs primarily in the cell cytoplasm. Here acetyl CoA is used to make palmitoyl CoA. Extension of the carbon chain to stearate and desaturation to form carbon-carbon double bonds occurs in mitochondria and the endoplasmic reticulum.
(b) Fatty acid catabolism does not occur in the same location as anabolism. The enzymes for β oxidation are located in the mitochondrial matrix.

29.17 In fatty acid synthesis, the compound that is added repeatedly to the enzyme, fatty acid synthase, is malonyl CoA, which has a three-carbon chain.

29.19 The carbon dioxide is released from malonyl-ACP which leads to the addition of two carbons to the growing fatty acid chain.

29.21 If one considers only what is happening to the fatty acid, removal of two hydrogens and two electrons, then it looks like oxidation only. However, the reaction is much more complex and involves a cofactor, NADPH and the substrate, oxygen. Both the fatty acid and NADPH undergo two-electron oxidation. The four electrons and protons are used to reduce oxygen to water:

$$O_2 + 4H^+ + 4e^- \longrightarrow 2H_2O$$

29.23 The only structural difference between NADH and NADPH is a phosphate group on one of the ribose units of NADPH. When considering the binding of NADPH to an NADH-requiring enzyme, two factors are important—size and charge. The phosphate makes the NADPH bulky and the NADH binding site may not be able to accommodate the larger size of the cofactor. In terms of charge, NADPH has two negative charges not present in NADH. The NADH binding site may have amino acid residues that have negatively-charged side chains like Glu or Asp. These would repel NADPH, but could hydrogen bond to the free hydroxyl group in NADH.

29.25 Humans have enzymes that catalyze the oxidation of saturated fatty acids to mono-unsaturated fatty acids with the double bond between carbons 9 and 10. For example, we can make palmitoleic acid from palmitic acid and oleic acid from stearic acid. Humans do not have enzymes that introduce a double bond beyond the 10th carbon. Therefore humans cannot make linoleic (double bonds at carbons 9-10 and 12-13) or linolenic acid (double bonds at carbons 9-10, 12-13, and 15-16). Those fatty acids are essential in the diet.

29.27 To make a glucoceramide, sphingosine is reacted with an acyl CoA that adds a fatty acid in amide linkage. Glucose is added to the hydroxyl group of sphingosine using the activated form, UDP-glucose (see Section 21.8).

29.29 All of the carbons in cholesterol orginate in acetyl CoA. An important intermediate in the synthesis of the steroid is a C-5 fragment called isopentenyl pyrophosphate:

$$\underset{\text{C-2}}{\text{3AcetylCoA}} \longrightarrow \underset{\text{C-6}}{\text{mevalonate}} \longrightarrow \underset{\text{C-5}}{\text{isopentenyl pyrophosphate}} + CO_2$$

29.31 An amino acid is synthesized by the reverse of oxidative deamination (Section 29.5). The amino acid product is aspartate. NADH will be oxidized to NAD^+.

29.33 The products of the transamination reaction shown are valine and α-ketoglutarate.

$$(CH_3)_2CH\text{-}\overset{O}{\overset{\|}{C}}\text{-}COO^- + {}^-OOC\text{-}CH_2\text{-}CH_2\text{-}\underset{NH_3^+}{\underset{|}{C}H}\text{-}COO^- \longrightarrow$$

The keto form of valine — Glutamate

$$(CH_3)_2CH\text{-}\underset{NH_3^+}{\underset{|}{C}H}\text{-}COO^- + {}^-OOC\text{-}CH_2\text{-}CH_2\text{-}\overset{O}{\overset{\|}{C}}\text{-}COO^-$$

Valine — α-Ketoglutarate

29.35 The carbon dioxide that is used to make carbohydrates in plants is reduced by the cofactor NADPH.

29.37 (a) The colored urine of blue diaper syndrome is caused by indigo blue dye.
(b) It is formed from the oxidation of the amino acid tryptophan.

29.39 The bonds that connect the nitrogen bases to the ribose units are β-N-glycosidic bonds just like those found in nucleotides.

29.41 The amino acid produced by transfer of the amino group is Phe.

29.43 The structure of a lecithin (also phosphatidyl choline) is in Section 21.6. Its synthesis requires activated glycerol, two activated fatty acids, and activated choline. Since each activation requires one ATP, the total number of ATP molecules needed is four.

29.45 The compound that reacts with Glu in a transamination reaction to form serine is 3-hydroxypyruvate. The the reaction is shown below:

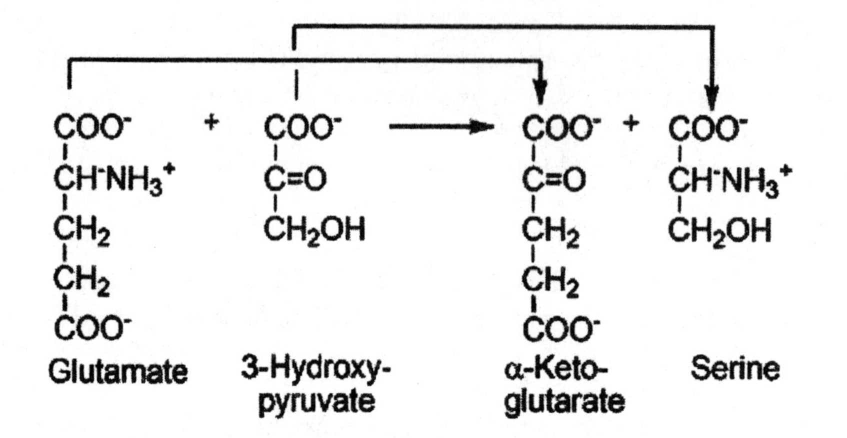

29.47 HMG-CoA is 3-hydroxy 3-methylglutaryl-CoA. Its structure is shown in Section 29.4. Carbon 1 is the carbonyl group linked to the thiol group of CoA.

29.49 Heme is a porphyrin ring with an iron ion at the center. Chlorophyll is a porphyrin ring with a magnesium ion at the center.

29.51 Fatty acid biosynthesis takes place in the cytoplasm, requires NADPH, and uses malonyl-CoA. Fatty acid catabolism takes place in the mitochondrial matrix, produces NADH and $FADH_2$, and has no requirement for malonyl-CoA.

29.53 Photosynthesis has high requirements for light energy from the sun.

29.55 Lack of essential amino acids would hinder the synthesis of the protein part. Gluconeogenesis can produce sugars even under starvation conditions.

29.57 Separation of catabolic and anabolic pathways allows for greater efficiency, especially in control of the pathways. It allows for both processes to be going on simultaneously in the body as conditions can be different in different cell compartments.

Chapter 30 Nutrition

30.1 No, nutrient requirements vary from person to person. The recommended daily allowances (RDA) stated by the government are average values and they assume a wide range of needs in different individuals. Some of the factors that change dietary needs include age, level of activity, genetics, and geographical region of residence.

30.3 Sodium benzoate is not catabolized by the body; therefore, it does not comply with the definition of a nutrient—components of food that provide growth, replacement, and energy. Calcium propionate enters mainstream metabolism by conversion to succinyl CoA and catabolism by the citric acid cycle.

30.5 The Nutrition Facts label found on all foods must list the percentage of daily values for four important nutrients: vitamins A and C, calcium, and iron.

30.7 Fiber is an important non-nutrient found in some foods. It is the indigestible portion of fruits, vegetables, and grains. Chemically, fiber is cellulose, a polysaccharide that cannot be degraded by humans. It is important for proper operation of dietary processes, especially in the colon.

30.9 The basal caloric requirement is calculated assuming the body is completely at rest. Because most of us perform some activity, we need more calories than this basic minimum.The caloric intake of 2100 for a young woman is a peak requirement. The difference between the basic level and the peak is needed to produce energy for activity.

30.11 Assume that each pound of body fat is equivalent to 3500 Cal. Therefore, the total number of calories that must be deleted from the diet is 3500 Cal/lb x 20 lb = 70,000 Cal. Because this must be done in 60 days, the amount restricted each day is 70,000 Cal ÷ 60 days = 1167 Cal/day. The caloric intake each day should be 3000 – 1167 = 1833 Cal, or approximately 1800 Cal/day.

30.13 Water makes up about 60 % of our body weight; therefore, one might assume she/he can lose weight by taking diuretics to enhance water release. However, this would only serve as a "quick fix" as the weight would return rapidly. In addition, it could be dangerous because the body needs a constant supply of water. The only way to effectively lose weight is to reduce the level of body fat.

30.15 Amylose is a storage polysaccharide that is one of the components in starch. Chemically it is an unbranched polysaccharide composed of glucose residues linked by α-1,4-glycosidic bonds. α-Amylase is an enzyme that catalyzes the hydrolysis of the glycosidic bonds at random sites in amylose. Thus the product would be different-sized, oligosaccharide fragments much smaller than the original amylose molecules.

30.17 No. Dietary maltose, the disaccharide composed of glucose units linked by an α-1,4-glycosidic bond, is rapidly hydrolyzed in the stomach and small intestines. By the time it reaches the blood, it is the monosaccharide glucose.

30.19 Arachidonic acid is produced from the essential, unsaturated fatty acid, linoleic acid.

30.21 No, lipases degrade neither cholesterol nor fatty acids. Lipases catalyze the hydrolysis of the ester bonds in triglycerides, releasing free fatty acids and glycerol.

30.23 Yes, it is possible for a vegetarian to obtain a sufficient supply of adequate proteins; however, the person must be very knowledgeable about the amino acid content of vegetables. It is very difficult to find a single vegetable that has complete proteins, that is, proteins with every essential amino acid. It is important for a vegetarian to eat a variety of vegetables in a proper balance so that all of the essential amino acids, fats, and nutrients are present.

30.25 Dietary proteins begin degradation in the stomach that contains HCl in a concentration of about 0.5 %. Trypsin is a protease present in the small intestines that continues protein digestion after the stomach. Stomach HCl denatures dietary protein and causes somewhat random hydrolysis of the amide bonds in the protein. Fragments of the protein are produced. Trypsin catalyzes hydrolysis of peptide bonds only on the carboxyl side of the amino acids Arg and Lys.

30.27 The rice/water diet provides sufficient calories for the basal caloric requirement; however, the diet is lacking important nutrients such as some essential amino acids (e.g.,Thr, Lys), essential fats, vitamins, and minerals. It is expected that many of the prisoners will develop deficiency diseases in the near future.

30.29 Limes provided sailors with a supply of vitamin C to prevent scurvy.

30.31 Vitamin K is essential for proper blood clotting.

30.33 The only disease that has been proven scientifically to be prevented by vitamin C is scurvy.

30.35 Vitamins E and C, and the carotenoids may have significant effects on respiratory health. This may be due to their activity as antioxidants.

30.37 There is a sulfur atom in biotin and in vitamin B-1 (also called thiamine).

30.39 The original food pyramid did not consider the difference between types of nutrients. It assumed that all fats were to be limited and that all carbohydrates were healthy. The new guide recognizes that polyunsaturated fats are necessary and that carbohydrates from whole grains are better for you than those from refined sources.

30.41 All proteins, carbohydrates, and fats in excess have metabolic pathways that lead to increased levels of fatty acids. However, there is no pathway that allows fats to generate a net surplus of carbohydrates. Thus, fat stores cannot be used to make carbohydrates when a person's blood glucose is low. This reality has a lot to do with the difficulty of diets.

30.43 All effective weight loss is based upon increasing activity while limiting caloric intake. However, it is more effective to concentrate on increasing activity than limiting intake. While it is difficult to lose weight by dieting because of the fact that fats cannot be used to make carbohydrates, exercise uses up fatty acids quickly without necessarily depleting glycogen stores. This makes exercise a more efficient way to lose weight.

30.45 (a) Most studies show that the artificial sweeteners Sucralose and acesulfame-K are not metabolized in any measurable amounts; they pass through unchanged.
(b) Digestion of aspartame can lead to high levels of phenylalanine, which many people are sensitive to.

30.47 Note the structure of creatine in Chemical Connections 30D. The amino acid that most resembles the structure of creatine is arginine. The top part of creatine, the carbon bound to three nitrogens, is the same as the side chain of Arg. Also, both have carboxylate groups at the other end of the molecule.

30.49 Carbohydrates should be consumed before the event as part of a high-carbohydrate diet. This is still considered the best diet for athletes. For competitions lasting longer than about an hour, carbohydrates should also be consumed during the event to prevent the complete depletion of glycogen and subsequent lowering of blood glucose levels.

30.51 Caffeine acts as a central nervous system stimulant, which provides a feeling of energy that athletes often enjoy. In addition, caffeine reduces insulin levels and stimulates oxidation of fatty acids, which would be beneficial to endurance athletes.

30.53 The vitamin pantothenic acid is part of Coenzyme A.
(a) Glycolysis: pyruvate dehydrogenase in Reaction 12, Figure 28.4, uses Coenzyme A.
(b) Fatty acid synthesis: the first step that involves the enzyme fatty acid synthase.

30.55 Proteins that are ingested in the diet are degraded to free amino acids that are then used to build proteins that carry out specific functions. Two very important functions include structural integrity and biological catalysis. Our proteins are constantly being turned over, that is, continuously being degraded and rebuilt using free amino acids.

30.57 The very tip of the food pyramid displays fats, oils, and sweets, with the cautionary statement, “Use sparingly”. We can omit sweets, meaning refined sugars, completely from the diet; however, complete omission of fats and oils is dangerous. We must have dietary fats and oils that contain the two essential fatty acids, linoleic acid and linolenic acid. It may be possible that the essential fatty acids are present as components in other food groups, i.e. the meat, poultry, fish group.

30.59 Walnuts are not just a tasty snack, they are a healthy one. Walnuts have protein, in fact, nuts are included in a group of the Dept. of Agriculture’s food pyramid. Walnuts are also a good source of vitamins and minerals including vitamins E, B, biotin, potassium, phosphorus, zinc, manganese, and others.

30.61 No, the lecithin is degraded in the stomach and intestines long before it could get into the blood. The phosphoglyceride is degraded to fatty acids, glycerol, and choline, that are absorbed through the intestinal walls.

30.63 Patients who have had ulcer surgery are administered digestive enzymes that may have been lost during the procedure. The enzyme supplement should contain proteases to help break down proteins, and lipases to assist in fat digestion.

Chapter 31 Immunochemistry

31.1 Examples of external innate immunity include action by the skin, tears, and mucus, all of which are barriers to entrance of foreign particles.

31.3 The skin fights infection by providing a barrier against penetration of pathogens. The skin also secretes lactic acid and fatty acids, both of which create a low pH thus inhibiting bacterial growth.

31.5 Innate immunity processes have little ability to change in response to immune dangers. The key features of adaptive (acquired) immunity are specificity and memory. The acquired immune system uses antibody molecules designed for each type of invader. In a second encounter with the same danger, the response is more rapid and more prolonged than the first.

31.7 T cells originate in the bone marrow, but grow and develop in the thymus gland. B cells originate and grow in the bone marrow.

31.9 Macrophages are the first cells in the blood that encounter potential threats to the system. They attack virtually anything that is not recognized as part of the body including pathogens, cancer cells, and damaged tissue. Macrophages engulf an invading bacteria or virus and kill it using nitric oxide (NO, see Table 31.1).

31.11 T cells recognize only peptide/protein antigens. This is in contrast to B cells which can recognize antigens from other molecule types, such as complex carbohydrates.

31.13 Class II MHC molecules pick up damaged antigens. A targeted antigen is first processed in lysosomes where it is degraded by proteolytic enzymes. An enzyme, GILT, reduces the disulfide bridges of the antigen. The reduced peptide antigens unfold and are further degraded by proteases. The peptide fragments remaining serve as epitopes that are recognized by class II MHC molecules (Figure 31.7).

31.15 MHC molecules are transmembrane proteins that belong to the immunoglobulin superfamily. They have peptide-binding variable domains. Class I molecules are single-chain polypeptides, whereas class II are protein dimers. They are originally present inside cells until they become associated with antigens and move to the surface membrane.

31.17 If we assume that the rabbit has never been exposed to the antigen, the response will be 1-2 weeks after the injection of antigen.

31.19 (a) IgE molecules have a carbohydrate content of 10-12% which is equal to that of the IgM molecules. IgE molecules have the lowest concentration in the blood. The blood concentration of IgE is about 0.01-0.1 mg/100mL of blood.
(b) IgE molecules are attached to basophils and mast cells. When they encounter their

antigens, the cells release histamines that cause the effects of hay fever and other allergies. They are also involved in the body's defense against parasites.

31.21 The two F_{ab} fragments would be able to bind an antigen. Note in Figure 31.8 that these fragments contain the variable protein sequence regions and hence are able to be changed during synthesis against a specific antigen.

31.23 Immunoglobulin superfamily refers to all of the proteins that have the standard structure of a heavy chain and a light chain. In terms of amino acid sequence, each molecule has a constant region and a variable region. Molecules in the immunoglobulin superfamily are the Ig molecules themselves (e.g., IgG, IgM, IgE), T-cell receptors, and the MHCs.

31.25 Antibodies and antigens are held together by weak, noncovalent interactions. These are: hydrogen bonds, electrostatic interactions (dipole-dipole), and hydrophobic interactions. We have encountered these kinds of interactions also in enzyme-substrate complexes and hormone-receptor complexes.

31.27 Antibody diversity is caused by several factors. There are two major types of light chains, kappa and lambda, which can combine with heavy chains. Each type of chain shows the recombination of the V, J, and D genes discussed in 31.4D, when the B and T cells are differentiating. In addition, mutations occur as the cells are maturing, causing another level of diversity.

31.29 T cells carry on their surfaces unique receptor proteins that are specific for antigens. These receptors (TcR), members of the immunoglobulin superfamily, have constant and variable regions. They are anchored in the T cell membrane by hydrophobic interactions. They are not able to bind antigens alone, but they need additional protein molecules called cluster determinants that act as coreceptors. When TcR molecules combine with cluster determinant proteins, they form T-cell receptor complexes (TcR complex).

31.31 The components of the T cell receptor complexes are (1) accessory protein molecules called cluster determinants and (2) the T-cell receptor.

31.33 The adhesion molecule in the TcR complex that assists HIV infection is the cluster determinant 4 (CD4).

31.35 The CD4 and CD8 molecules act as adhesion molecules, helping the T cell receptor bind to the antigens and to help the T cell dock with the antigen presenting cell or B cell.

31.37 Cytokines are glycoproteins that interact with cytokine receptors on macrophages,_B cells, and T cells. They do not recognize and bind antigens.

31.39 Chemokines are a class of cytokines that send messages between cells. They attract leukocytes to the site of injury and bind to specific receptors on the leukocytes.

31.41 Cytokines are classified in several ways, but the main way is via their secondary structures. Cytokines such as interleukin-2 contain four helical segments. Another type such as tumor necrosis factor have only β-pleated sheets for a secondary structure. Another one, such as epidermal growth factor has a combination of α-helix and β-pleated sheet. A subgroup of cytokines, called chemokines, have 4 cysteine residues.

31.43 The T cells mature in the thymus gland. During maturation those cells that fail to interact with MHC and thus cannot respond to foreign antigens are eliminated by a special selection process. T cells that express receptors (TcR) that may interact with normal-self antigens are eliminated by the same selection process.

31.45 A signaling pathway that controls the maturation of B cells is the phosphorylation pathway activated by tyrosine kinase and deactivated by phosphatase.

31.47 The cytokines and chemokines are thought to play a major role in autoimmune diseases.

31.49 Human Immunodeficiency Virus binds to and enters the helper T cells in humans.

31.51 HIV is difficult to find because it mutates so quickly, making it difficult to create an effective vaccine against the virus. The gp120 protein on the virus changes conformations when it binds to the CD4 molecule, so antibodies elicited against the virus during the active part of the infection will not recognize the virus that is floating free. HIV also has proteins on its surface that block the action of natural killer cells, so it can evade the innate immune system. The virus also cloaks its outer membrane with sugars that are very similar to the natural sugars of the host cell membranes.

31.53 It is difficult to create an effective antibody to a virus that mutates quickly. Even if the antibody is successful at neutralizing most of the virus particles, just a few survivors can repopulate the host very quickly. In addition, most attempts to generate antibodies have led to large numbers of antibodies that were not effective.

31.55 It had been known since the 1860's that the mayapple had anticancer properties. It was later found that a chemical found in the mayapple, picropodophyllin, inhibits spindle formation during mitosis in dividing cells. As all chemotherapy agents do, they hinder rapidly dividing cells, like cancer cells, more than regular cells.

31.57 MCA treatment for cancer is better than chemotherapy because MCAs are much more specific for the cancer cells. Treatment is directed toward just the cancerous cells, not all cells in general like chemotherapy. In addition, MCA therapy shows minimal toxicity.

31.59 Jenner noticed that people infected with cowpox did not also get smallpox. He tried the experiment of infecting a boy with cowpox, a mild disease, and then later injecting the boy with smallpox. The boy developed no symptoms of the dangerous smallpox. Today this dangerous and unethical experiment could not be done by a reputable physician.

31.61 Vaccination is a French word that literally means, "encowment." It was a derisive term used by some scientists who did not believe in Jenner's theory.

31.63 Dendritic cells are the first to encounter pathogens, followed by macrophages and natural killer cells.

31.65 Chemokines (or more generally cytokines) help leukocytes migrate out of a blood vessel to the site of injury. Cytokines help the proliferation of leukocytes.

31.67 A compound called 12:13 dEpoB, a derivative of epothilon B, is being studied as a anticancer vaccine.

31.69 TNF (tumor necrosis factor) receptors are found on the surface of tumor cells.